ROUTLEDGE LIBRARY EDITIONS: ENERGY

Volume 10

ENERGY ISSUES AND OPTIONS FOR DEVELOPING COUNTRIES

ENERGY ISSUES AND OPTIONS FOR DEVELOPING COUNTRIES

UNITED NATIONS

Routledge
Taylor & Francis Group

LONDON AND NEW YORK

First published in 1989 by Taylor & Francis

This edition first published in 2019
by Routledge
2 Park Square, Milton Park, Abingdon, Oxon OX14 4RN

and by Routledge
52 Vanderbilt Avenue, New York, NY 10017

Routledge is an imprint of the Taylor & Francis Group, an informa business

British Library Cataloguing in Publication Data
A catalogue record for this book is available from the British Library

ISBN: 978-0-367-21122-6 (Set)
ISBN: 978-0-429-26565-5 (Set) (ebk)
ISBN: 978-0-367-21121-9 (Volume 10) (hbk)
ISBN: 978-0-367-21140-0 (Volume 10) (pbk)
ISBN: 978-0-429-26564-8 (Volume 10) (ebk)

Publisher's Note
The publisher has gone to great lengths to ensure the quality of this reprint but points out that some imperfections in the original copies may be apparent.

Disclaimer
The publisher has made every effort to trace copyright holders and would welcome correspondence from those they have been unable to trace.

Energy Issues and Options for Developing Countries

Published for and on behalf of the United Nations

Taylor & Francis

New York • Philadelphia • Washington D.C. • London

USA	Publishing Office:	Taylor & Francis · New York
		79 Madison Ave., New York, NY 10016
	Sales Office:	Taylor & Francis · Philadelphia
		242 Cherry St., Philadelphia, PA 19106-1906
UK		Taylor & Francis Ltd.
		4 John St., London WC1N 2ET

Energy Issues and Options for Developing Countries

Copyright © 1989 The United Nations

First published 1989
Printed in the United States of America

Library of Congress Cataloging in Publication Data

Energy issues and options for developing countries.
    p. cm.
  Bibliography: p.
  Includes index.
ISBN 0-8448-1578-0
   1. Power resources—Developing countries. 2. Energy policy-
-Developing countries. I. United Nations.
TJ163.25.D44E533 1989
333.79'09172'4—dc19

                              88-39591
                                 CIP

Contents

Part III. Small Energy Packages

Part IV. Application of Microcomputer Technology in Energy Planning

List of Figures

List of Tables

Preface

Since its creation in 1945, the United Nations has been committed to promoting economic and social progress in developing countries, and over the past four decades UN assistance has played an important role in supplementing these countries' own development efforts. UN technical cooperation has grown rapidly since the early days of the organization, reflecting not only an increase in the level of activity, but also far-reaching changes in its emphasis, as developing countries have progressed and as new strategies have emerged to guide development efforts.

The Department of Technical Co-operation for Development (DTCD) is the principal arm of the United Nations for technical cooperation activities. Among its recent activities, the department has helped developing countries discover minerals and oil, improve water supplies, prepare development plans, establish remote sensing centers, take censuses, and analyze changes in population distribution, install computers, strengthen public administration, build roads, draw maps, develop rural areas, and improve the status of women.

DTCD's efforts in the energy area span over two decades. The department has assembled considerable technical expertise in energy development in order to efficiently respond to requests for assistance, and is currently executing projects amounting to $97 million. Its work has encompassed:

- establishment of energy planning and information systems and development of programs for energy conservation
- exploration and development of hydrocarbon resources
- rehabilitation and modernization of electricity power plants
- exploration and development of geothermal resources and construction of geothermal power plants
- assessment of hydro resources, implementation of feasibility studies, and construction of small-scale hydropower plants
- establishment of solar, wind, and biomass demonstration projects
- strengthening of national energy institutions
- training program for scientists, engineers, technicians, and managers in the latest technologies and techniques in the field of energy

The role of DTCD in international technical cooperation is primarily one of technical assistance rather than of financing development projects. DTCD is currently executing 117 energy projects in 46 developing countries. The projects play an important catalytic role in many developing countries, especially at an early stage of their energy development effort. This applies in particular to strengthening of energy sector institutions through training, provision of experts, and acquisition of modern technology.

DTCD has promoted the international exchange of experiences in all aspects of energy development through the organization of meetings, symposia, and workshops on the following topics: oilfield development techniques; energy supply management; coal for electricity generation; small-scale hydropower; microcomputer software applications for energy planning; new technologies for the development of coal resources; economics of small renewable energy systems; development and exploitation of geothermal energy; increasing the reliability and efficiency of electricity generation, transmission and distribution; rural energies and rural electrification; and solar heating and cooling.

DTCD engages in research in support of technical cooperation activities. In the energy field, research focuses on leading edge technologies. The results of this research are disseminated in the form of technical papers describing recent trends or new technologies and methods. The four technical studies in this volume represent a selection of recently completed work.

Part I. Enhanced Oil Recovery. Enhanced oil recovery (EOR) is an issue that will continue to grow in importance as oil reserves dwindle. Oil remaining in reservoirs after primary and secondary production operations is the target for enhanced oil recovery. Even in developed countries where primary and secondary production methods are quite mature, a large percentage (commonly 30–90%) of the oil in a reservoir gets left in the ground. In developing countries where production techniques may not be as sophisticated, recovery efficiencies are lower.

A chief advantage of EOR is that there are no exploration costs since this oil exists in known reservoirs and, in some cases, existing wells and surface equipment can be used. In very few, if any, cases will reservoirs be redeveloped for EOR once the wells have been plugged and abandoned, as additional costs would be prohibitive. Loss of potential resource through abandonment should bring a sense of urgency to EOR development throughout the world, especially in developing countries.

This section provides an overview of the various EOR processes and reviews their application. The state-of-the-art as developed in the United States and worldwide is discussed relating technical and economic and other considerations that influence EOR development. Finally, the section lists those organizations and institutions where opportunities for training and technology transfer are available.

Part II. Small-Scale Hydropower Programs. Since the early 1970s, the economic attractiveness of small-scale hydropower has greatly increased in developing countries where the institutional and financial situation and the lack of adequately large power markets tend to inhibit the development of main river schemes. With the knowledge that rural electrification stimulates economic progress in rural and outlying areas, many developing countries are examining their water resources to determine how best they might be exploited. The size of a scheme considered as "small scale" in this section would lie between 500 kW and 10 MW, but the bounds of this range are arbitrarily chosen.

The benefits from a local hydro scheme do not necessarily accrue to the local electricity consumer; the scheme can bring wider benefits for the national economy as a whole by:

- reducing the drain on foreign currency reserved where hydropower can replace diesel power
- reducing deforestation where the electricity produced can replace firewood
- reducing the division between urban and rural life and the social stresses this produces
- discouraging population drift to urban areas through the improvement of living conditions of the rural population

This section goes on to discuss all aspects of small-scale hydropower planning, economics and financing, development, operation, management, and training. It is a thorough treatment of the subject and should serve as a valuable reference tool and management guide for hydropower development for several years to come.

Part III. Small Energy Packages. The majority of the world's population depends on small internal combustion engines or engine generators to provide electricity because they live in areas too distant from the central power grid to receive electric service. The cost of the petroleum products that these engines consume and the skilled maintenance that they require are a considerable burden to oil-importing developing countries. Many of these engines and generators are candidates for eventual replacement by new and renewable sources of energy. Although the development and diffusion of small renewable energy sources based on solar energy, wind energy, hydropower, or biofuels have been slower than expected, nevertheless, the technologies and efficiencies have been improving while costs have fallen.

Generally speaking, the energy source of choice, if it were available at a competitive cost, would undoubtedly be electricity. The delivered cost of electric energy in the rural areas to an extended grid is the reference against which other alternatives of decentralized energy production must be compared. But the viability of these alternatives cannot be assessed by this comparison alone.

Modern techniques for the utilization of renewable energies make it possible to:

- Supply a mini or micro-distribution grid by a power station of comparable size
- Satisfy a local energy need, such as lighting, refrigeration, water pumping, cooking, telecommunications, etc.

This section on small energy packages examines all existing and potential energy sources and their applications in the rural environment. Considered are internal combustion engines, micro-hydropower stations, photovoltaic power stations, wind electricity generators, electricity generation from biomass, solar and wind-powered water pumping, as well as renewable energy applications for domestic energy needs, rural telecommunications, and health centers. Each energy package is described in detail by a general definition, a technological state-of-the-art, constraints to widespread diffusion, costs, potential for future wide-scale development, and advantages and disadvantages to its application.

Part IV. Application of Microcomputer Technology in Energy Planning. After the oil price shocks of the 1970s, the flurry of analysis of the energy market centered on the impact of higher oil prices on balance of payments. Energy conservation and development of domestic resources were primary objectives. In the 1980s, however, other serious issues in energy planning have emerged. They include fuelwood use and the relationship between energy and the broader aspects of rural development and agricultural policies. Another concern is the deterioration of the international financial environment with many developing countries experiencing critical debt re-payment difficulties. Problems of capital mobilization have made investment planning much more difficult and have given rise to calls from the IMF and World Bank for reforms in energy pricing and tariff structures. None of these issues is likely to disappear quickly. They dictate that approaches to energy planning and policy analysis must be more sophisticated and must deal with a broader set of developmental and financial constraints.

Along with the need for improved analytical approaches has appeared the rapid development of microcomputer hardware and software. Thus microcomputers have become a natural adjunct to energy analysis since the types of analysis required lend themselves well to the microcomputer environment. However, the scope, appropriateness, and sophistication of microcomputer energy models vary quite considerably, and their usefulness in any given situation is dependent on a variety of factors.

This section is designed to serve as a guide to planners and analysts on the subject of microcomputer-based energy planning. It provides:

- criteria by which models should be designed and selected
- critical reviews of available software
- issues in hardware procurement
- information on current hardware costs
- lessons to be drawn from case studies in developing countries

Together the four parts included in this book present a sampling of the wide range and scope of technological development on-going in the field of energy at this time. Wherever possible, an effort has been made to include case studies of the application of these technologies in developing countries. In this way DTCD hopes to interest governments in evaluating the appropriateness of these technologies to their specific energy needs. If DTCD can then assist countries in acquiring and implementing such technologies so that they contribute in a sustainable manner to overall development, DTCD will have fulfilled its role.

Part I

Enhanced Oil Recovery

1

Introduction

Large volumes of oil remain in reservoirs after applying primary and secondary production methods. In the more mature oil-producing areas of the world such as the United States and some countries of western Europe, a great deal of effort has gone into improving the recovery efficiency of primary and secondary operations. Even so, these methods of operation still leave more than half of the oil present in the reservoirs at the time of discovery. In the developing countries of the world, the primary and secondary methods of operation are sometimes not too sophisticated. Consequently, recovery efficiencies are not as high as in the United States and other more developed countries. The oil remaining in the reservoirs of these countries after primary and secondary operations may be significantly more than half of the in-place oil at the time of discovery. Also, some of the reservoirs of the developing countries have been exploited through recovery by "skimming." Here, only the most easily recovered oil is produced from the reservoir and a significant fraction of the original oil in place remains in the reservoir.

Oil remaining in reservoirs after primary and secondary operations is the target for enhanced oil recovery (EOR) operations. Vast quantities of this kind of target oil are available in reservoirs of the developing countries. This oil is located in known reservoirs and represents a national resource base for each of the developing countries. No additional funds are required to find the oil. In some cases, existing wells and surface equipment can be used to reduce the costs of EOR operations. Of course, existing equipment becomes less valuable with time as normal wear by corrosion and other factors takes its toll.

Another and perhaps the most important cost factor relates to the oil saturation existing in the reservoir at the start of EOR. Generally, the higher the reservoir oil saturation at the start of EOR operation, the lower the per barrel cost for recovery. Here, the developing countries should be in a good position because their reservoirs normally would have higher oil saturation at the completion of the conventional recovery operations.

In the developing countries, EOR should be looked upon as a method of recovering an otherwise unrecoverable national resource. Also, EOR should be considered an alternate liquid hydrocarbon source when considering the development of other liquid hydrocarbons sources such as oil shale, tar sands, and coal. In most cases, the EOR source will prove to be less expensive when compared on a cost basis with these other sources. However, the cost of EOR oil generally will be more than the oil recovered by more conventional methods. It should be kept in mind that: (1) conventional methods generally "skim the cream," (2) "skimming the cream" is the least expensive method of recovery, and (3) oil recovered by EOR can be obtained by no other method. The oil remaining in the reservoir is either recovered by EOR or may be lost forever once the reservoir is plugged and abandoned. In very few, if any, cases will reservoirs by redeveloped for EOR once the wells have been plugged and abandoned. In most cases, the additional costs will be prohibitive. Loss of potential resource through plugging and abandonment should bring a sense of urgency to EOR development throughout the world, especially in the developing countries.

This section starts with an overview of the different types of EOR processes including comments on screening criteria and injectant availiability. Next, application of EOR techniques is considered in terms of reservoir screening, reservoir evaluation, performance prediction and selected field histories. Then, EOR state-of-the-art is discussed on the basis of the status of EOR techniques in the United States and worldwide, as well as technical factors, economic considerations, and factors influencing EOR development. Finally, information is presented on EOR training along the lines of technology transfer and available training programs. Conclusions are given at the end of the section.

2

Overview

This section considers only the EOR processes categorized as chemical, miscible, and thermal. Immiscible is treated as a subset of miscible. Other types of EOR methods are being studied in the laboratory, such as electrical heating and injection of microorganisms. Commerical use of these methods appears to be far in the future, and it is not considered here.

Following an overview of the different types of chemical, miscible, and thermal EOR processes, the concept of screening criteria is introduced in terms of the need for the use of these criteria. Then, injectant availability is considered as affecting the selection and development of a particular EOR process.

TYPES OF EOR PROCESSES

Reduced interfacial tension and mobility control are the two factors that are used in EOR to improve oil recovery. Cleaning solvent will remove a spot of oil from cloth, whereas water will not. The solvent reduces or eliminates the interfacial tension, whereas water does not mix with the oil. Molasses pushes water from a tube effectively with its favorable mobility ratio, whereas water is not efficient in displacing molasses from the same tube.

Chemical methods use chemicals to lower the interfacial tension and polymer for mobility control. Miscible methods eliminate or reduce interfacial tension by developing a fluid mixture that is miscible with the in-place oil and injected gas. Thermal methods use heat to reduce the oil viscosity, thereby providing improved mobility control conditions in the reservoirs.

EOR processes are categorized as chemical, miscible, or thermal. The chemical category includes the polymer, surfactant, and alkaline flooding processes. Figures 2.1–2.3 show schematic diagrams of these three processes (1). In the miscible category are included processes in which a gas is injected into the reservoir under conditions where miscibility is obtained between the gas and reservoir oil. Gases of choice include carbon dioxide, hydrocarbons such as methane, ethane, propane, and

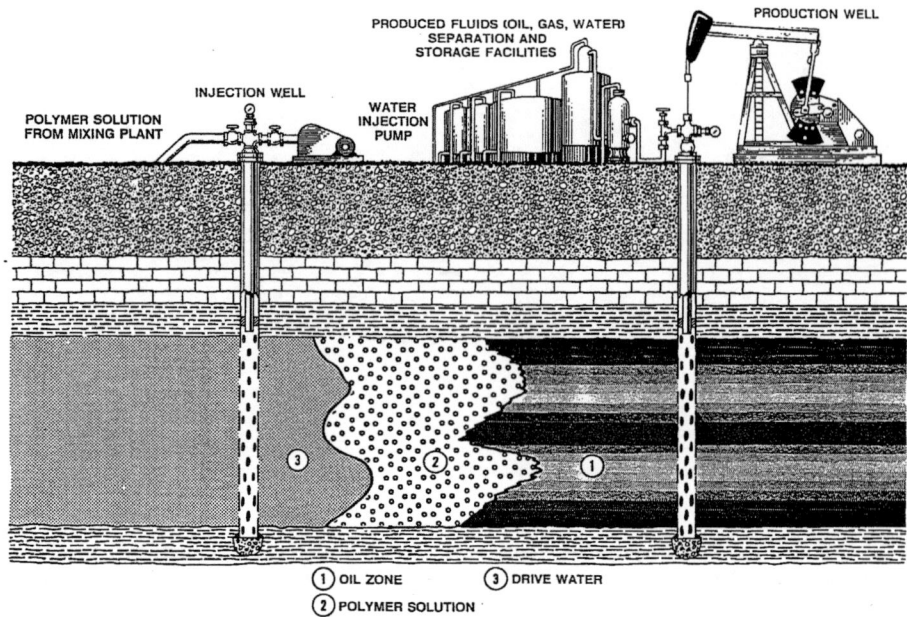

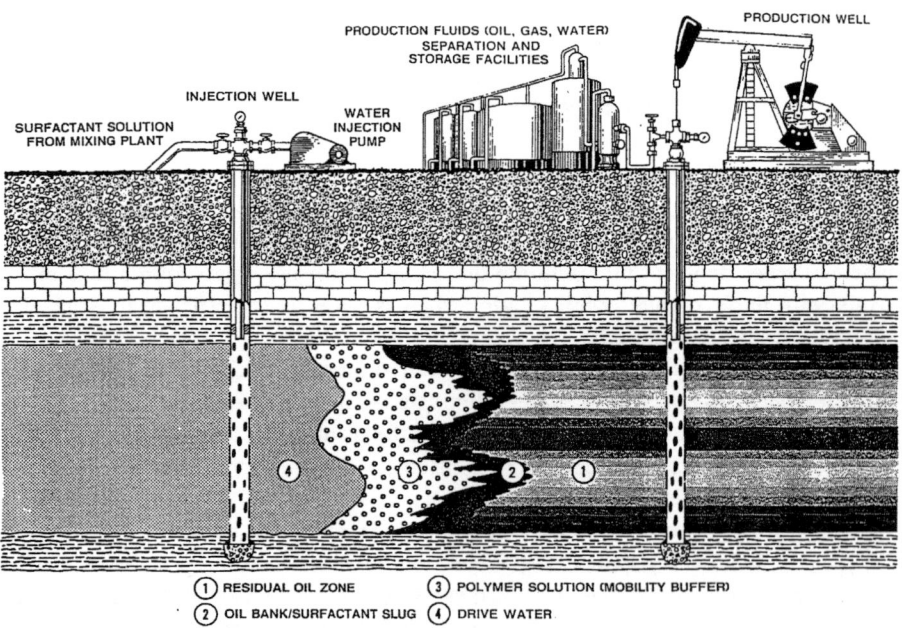

Figure 2.1 Polymer flooding. (*Source:* Adapted from original drawings by Joe R. Lindley, U.S. Department of Energy, Bartlesville Energy Technology Center)

Figure 2.2 Surfactant flooding. (*Source:* Adapted from original drawings by Joe R. Lindley, U.S. Department of Energy, Bartlesville Energy Technology Center)

others, and mixtures of these gases, and nitrogen. Figure 2.4 shows a schematic diagram for a miscible process where carbon dioxide is being injected. (1). The diagram would be similar for other injected gases. Steam and in-situ combustion processes are used in the thermal category. Cyclic and continuous steam injection are used in the steam process. Figures 2.5 and 2.6 illustrate the two techniques for steam injection (1). With the in situ combustion process, air is injected into the reservoir to provide oxygen to burn a portion of the oil. Water is sometimes simultaneously injected with air for mobility control. Figure 2.7 is a schematic of the wet in-situ combustion process (1).

Different types and amounts of chemicals are used in the three chemical processes. Both synthetic and biologically produced polymers have been used in polymer flooding, which differs from the other two chemical methods in that no reduction in interfacial tension takes place. Increased recovery by polymer flooding occurs because of improvement in vertical and areal conformance. As seen in Figure 2.2, a polymer solution also is used for mobility control behind the alkaline solution. Sometimes for mobility control polymer is mixed into the surfactant and alkaline solutions. At other times cost prohibits the use of polymer in either the surfactant of alkaline systems. In most of the surfactant systems, all chemicals are mixed at the surface before injection. A wide variety of surfactants have been used, including petroleum sulfonates, synthetic sulfonates and sulfates, and ethoxylated alcohols and sulfonates. The alkaline process differs from the surfactant process in that the surfactant is gen-

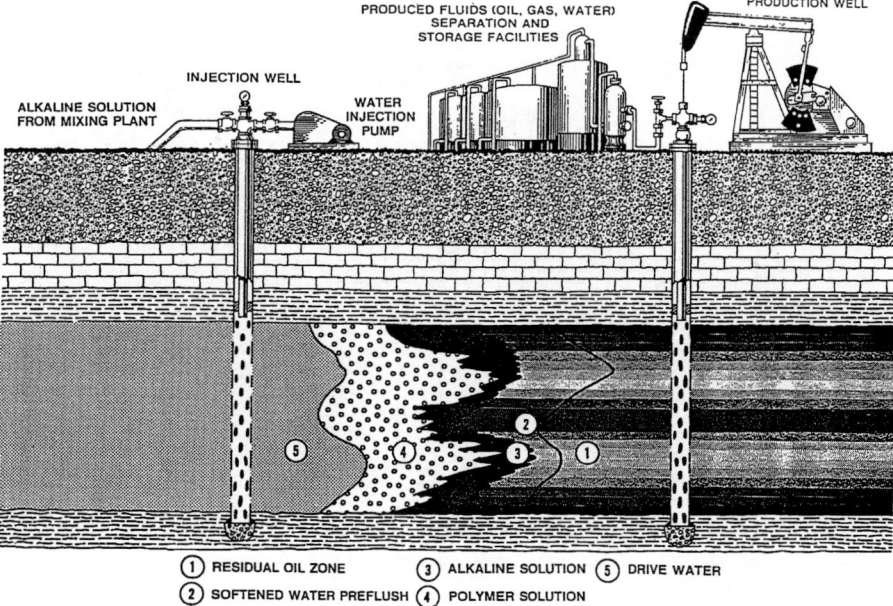

Figure 2.3 Alkaline flooding. (*Source:* Adapted from original drawings by Joe R. Lindley, U.S. Department of Energy, Bartlesville Energy Technology Center)

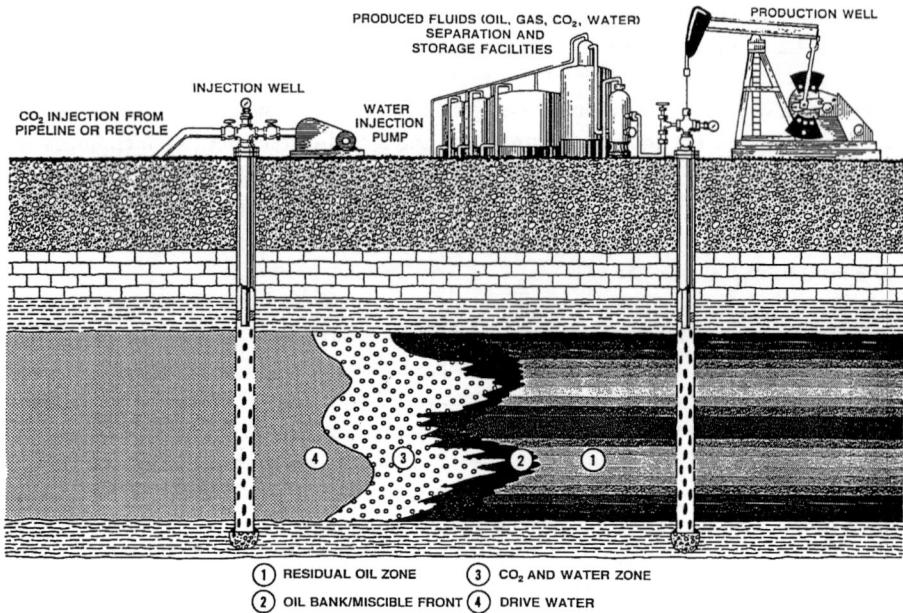

Figure 2.4 Miscible flooding. (*Source:* Adapted from original drawings by Joe R. Lindley, U.S. Department of Energy, Bartlesville Energy Technology Center)

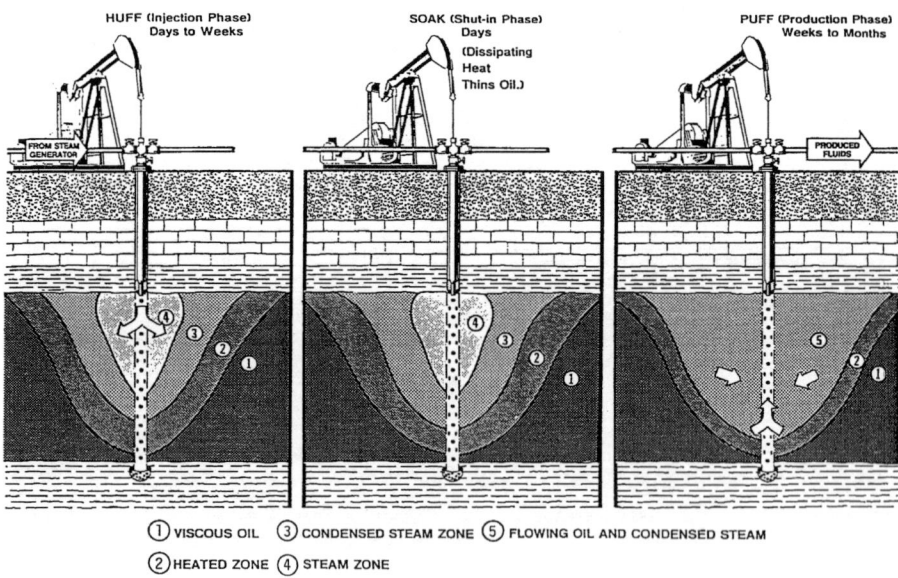

Figure 2.5 Cyclic steam stimulation. (*Source:* Adapted from original drawings by Joe R. Lindley, U.S. Department of Energy, Bartlesville Energy Technology Center)

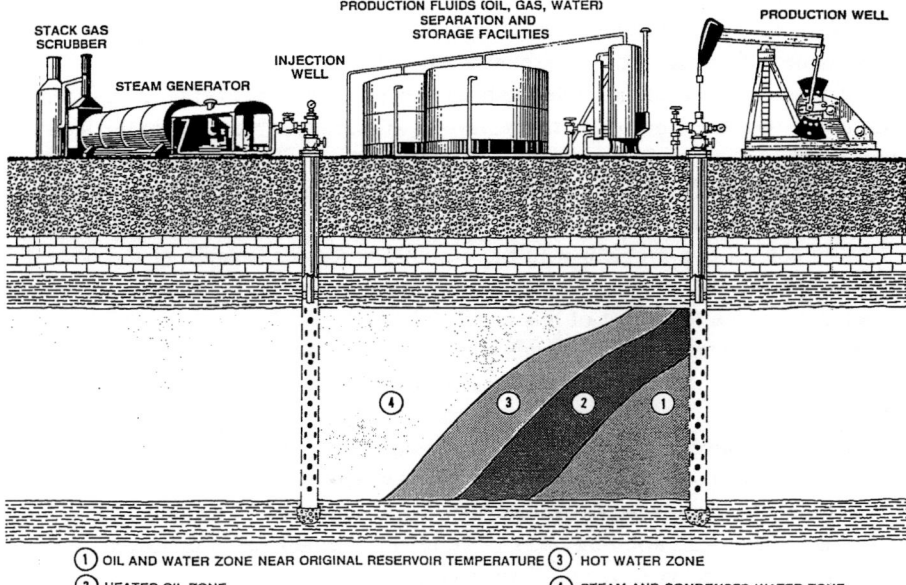

Figure 2.6 Steam flooding. (*Source:* Adapted from original drawings by Joe R. Lindley, U.S. Department of Energy, Bartlesville Energy Technology (Center)

① OIL AND WATER ZONE NEAR ORIGINAL RESERVOIR TEMPERATURE ③ HOT WATER ZONE
② HEATED OIL ZONE ④ STEAM AND CONDENSED WATER ZONE

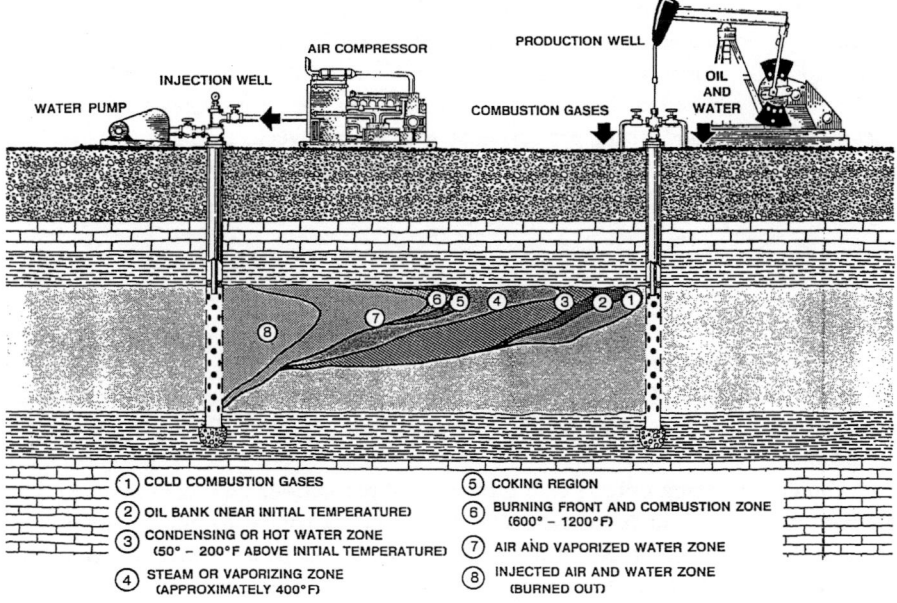

① COLD COMBUSTION GASES
② OIL BANK (NEAR INITIAL TEMPERATURE)
③ CONDENSING OR HOT WATER ZONE (50° – 200°F ABOVE INITIAL TEMPERATURE)
④ STEAM OR VAPORIZING ZONE (APPROXIMATELY 400°F)
⑤ COKING REGION
⑥ BURNING FRONT AND COMBUSTION ZONE (600° – 1200°F)
⑦ AIR AND VAPORIZED WATER ZONE
⑧ INJECTED AIR AND WATER ZONE (BURNED OUT)

Figure 2.7 Wet in-situ combustion. (*Source:* Adapted from original drawings by Joe R. Lindley, U.S. Department of Energy, Bartlesville Energy Technology Center)

erated in-situ. Here alkaline reacts with organic acids in the crude oil to form surfactant. Sodium hydroxide and various orthosilicates have been used as alkaline agents. Some experts feel that sodium carbonate has merit; others do not.

The gases used in the miscible methods generally do not experience first contact miscibility with the crude oil in the reservoir. Miscibility takes place in the reservoir after repeated contacts of a given volume of gas with fresh crude oil. Gas continues to extract light components from the fresh oil. After sufficient contacts, a bank is developed having a composition that is miscible with injected gas and fresh crude. Mobility control is a problem with the miscible processes because of the wide difference in viscosity between the injected gas and in-place crude. Water is sometimes injected to reduce the effect of this problem.

Reservoir pressures must be high enough at existing reservoir temperatures for miscibility to be achieved. If reservoirs are too shallow, miscibility may not be achievable. Carbon dioxide has a lower miscibility pressure than either nitrogen or methane. The miscibility pressure of methane can be reduced by adding heavier hydrocarbons such as propane and butane. Even where miscibility cannot be achieved, immiscible operations may be worthwhile. With immiscible operations, the oil viscosity is reduced and swelling takes place as the injected gas is dissolved. Both of these factors can provide increased recovery.

Thermal methods normally are used in reservoirs containing oil with higher viscosity. Steam injection accounts for over 80% of the ongoing thermal projects. In most thermal candidate reservoirs, primary and secondary recovery, if any, has been low and the remaining oil saturation is high. The high starting value of oil saturation frequently produces more favorable economics. This is one of the main reasons that of all EOR processes, thermal methods are more frequently found to be applied throughout the world. Cyclic steam injection is most often used to test the applicability of a prospect to continuous steam injection. Gravity override is one of the factors in continuous steam projects that reduces recovery. Surfactants have been and are being developed so as to remain stable at high temperatures. The injected surfactants create in-place emulsions that reduce the gravity override problem.

SCREENING CRITERIA

Screening criteria are used to determine the applicability of a particular EOR process for a given reservoir. Generally, these screening criteria are based on the technical aspects of the problem. With some reservoirs, more than one EOR process may be applicable.

Technical screening criteria take into account the various reservoir parameters including rock and fluid properties. Other criteria then have to come into play. A criterion may be used based on the total amount of oil to be recovered. Here, this oil may be important to the national interest. In this case, economics may or may not be used to make a final selection. For example, both the polymer and surfactant flooding processes may be applicable to the same reservoir. Surfactant flooding might recover more than four times as much oil as polymer flooding. But the cost per barrel of oil recovery with polymer flooding will be significantly less.

Different screening criteria have been developed from EOR results reported in the published literature. Generally, these criteria are based on field results. Tables summarizing some of these screening criteria are given in the literature (1,2,3,4).

Perhaps the most up-to-date criteria are given in the 1984 NPC study (1). These criteria were based on input from a large number of experts. Tables 2.1 and 2.2 show screening criteria for the implemented and advanced technology cases of this study (1). In preparing the implemented criteria, the experts used only those conditions where a technically successful field project for a particular EOR process had been completed and reported in the literature. All the different parameters given for a particular process were not obtained from the same field test. For example, the surfactant flooding <100,000 ppm implemented value for the salinity of formation brine is from the Louden test in Illinois, whereas the <40 cp in-situ oil viscosity is from the Wilmington test in California. The screening parameters for the implemented technology case give an idea of where the experts feel the various EOR processes stand at the present time. The advanced technology case represents a consensus by the experts as to what advances might occur by at least 1995.

The technical screening criteria given in Tables 2.1 and 2.2 can be used on a worldwide basis. Reservoirs in a developing country are no different from those in other parts of the world. In this section, any mention of screening criteria indicates those given in either Table 2.1 or 2.2.

INJECTANT AVAILABILITY

The availability or nonavailability of injectants either provides an incentive or causes a lack of incentive for using a particular EOR process. If injectants are not available within a country and EOR projects are to be started on a large scale in the future, plans should include providing an in-country means of obtaining the injectants. Pilot projects can be conducted using imported injectants. Full-scale commerical projects will be economically hampered unless injectants are available within a country. Two possible exceptions may be found in polymer and steamflooding. Exporting countries with a large polymer manufacturing capacity, such as Japan, may be willing to supply polymer to the developing countries at a reasonable price. In most steamfloods, low pressure boilers are used to provide the injected steam. Generally produced oil is used for fuel. Large-scale steamfloods may be possible with imported boilers.

The type of injectants available in some cases will provide guidance as to the type of EOR process to consider. Some countries have large deposits of inorganic salts such as sodium carbonate. The availability of this or other similar chemicals might indicate that alkaline flooding should be pursued. No natural carbon dioxide reservoir exists in some countries, such as India. If miscible flooding with carbon dioxide is

Table 2.1
Screening Criteria for EOR Candidates Implemented Technology Case[*]

Screening parameters	Units	Chemical flooding			Miscible flooding (carbon dioxide)	Thermal recovery	
		Surfactant	Polymer	Alkaline		Steam	In situ combustion
Oil gravity	°API	—	—	<30	≥25	10 to 34	10 to 35
In situ viscosity (μ)	cp	<40	<100	<90	—	≤15,000	≤5,000
Depth (D)	Feed	—	—	—	—	≤3,000	≤11,500
Pay zone thickness (h)	Feed	—	—	—	—	≥20	≥20
Reservoir temperature (T_R)	°F	<200	<200	<200	—	—	—
Porosity (ϕ)	Fraction	—	—	—	—	≥0.20[‡]	≥0.20[‡]
Permeability, average (k)	md	>40	>20	>20	—	250	≥35
Transmissibility (kh/μ)	md-ft/cp	—	—	—	—	≥5	≥5
Reservoir pressure (P_R)	psi	—	—	—	≥MMP[†]	≤1,500	≤2,000
Minimum oil content at start of process ($S_o \times \phi$)	Fraction	—	—	—	—	≥0.10	≥0.08
Salinity of formation brine (TDS)	ppm	<100,000	<100,000	<100,000	—	—	—
Rock type	—	Sandstone	Sandstone or Carbonate	Sandstone	Sandstone or Carbonate	Sandstone or Carbonate	Sandstone or Carbonate

[*]Other criteria of a geological and depositional nature were also considered. Generally, reservoirs with extensive faulting, lateral discontinuities, fractures, or overlying gas caps are not prime candidates for field-wide EOR application. These factors were considered during the manual screening step when they could be identified.
[†]MMP denotes minimum miscibility pressure, which depends on temperature and crude oil composition.
[‡]Ignored if oil saturation (S_o) × porosity (ϕ) criteria are satisfied.

Table 2.2
Screening Critiera for EOR Candidates Advanced Technology Case

Screening parameters	Units	Chemical flooding			Miscible flooding (carbon dioxide)	Thermal recovery	
		Surfactant	Polymer	Alkaline		Steam	In situ combustion
Oil gravity	°API	—	—	<30	≥25	—	—
In situ oil viscosity (μ)	cp	<100	<150	<100	—	—	≤5,000
Depth (D)	Feet	—	—	—	—	≤5,000	—
Pay zone thickness (h)	Feet	—	—	—	—	≥15	≥10
Reservoir temperature (T_R)	°F	<250	<250	<200	—	—	—
Porosity (ϕ)	Fraction	—	—	—	—	≥0.15[‡]	≥0.15[‡]
Permeability, average (k)	md	>10	>10	>10	—	≥10	≥10
Transmissibility (kh/μ)	md-ft/cp	—	—	—	—	—	—
Reservoir pressure (P_R)	psi	—	—	—	≥MMP[†]	≥2,000	≥4,000
Minimum oil content at start of process ($S_O \times \phi$)	Fraction	—	—	—	—	≥0.08	≥0.08
Salinity of formation brine (TDS)	ppm	<200,000	<200,000	<200,000	—	—	—
Rock type	—	Sandstone or Carbonate	Sandstone or Carbonate	Sandstone	Sandstone or Carbonate	Sandstone or Carbonate	Sandstone or Carbonate

Other criteria of a geological and depositional nature were also considered. Generally, reservoirs with extensive faulting, lateral discontinuities, fractures, or overlying gas caps were not prime candidates for held wide EOR application. These factors were considered during the manual screening step when they could be identified.

[†] MMP denotes minimum miscibility pressure, which depends on temperature and crude oil composition.

[‡] if oil saturation (S_O) × porosity (ϕ) criteria are satisfied.

to develop in these countries, source development would need to start obtaining this injectant from power or fertilizer plants. Surfactant flooding requires the use of chemical manufactured by using the most sophisticated technology. This fact should be considered if surfactant flooding is being considered for large-scale field development.

3

Application of EOR Techniques

Selection of a particular EOR process for a given reservoir requires the use of the most sophisticated reservoir engineering. The high cost of an EOR project justifies the need for complete geological and engineering studies before development starts. Limited money and resources initially spent on these studies will prevent conducting projects where failures will result in loss of large amounts of money. This principle of making initial studies is particularly important in developing countries where resources are limited. Initial studies should take place both in the laboratory and field. EOR expertise in the developing country should be supplemented by outside consultants either through assistance and/or by training. The initial study phase of an EOR project lends itself to an excellent opportunity to provide on-the-job training of the engineers and other technical people of a developing country. A part of the initial study phase includes evaluation of EOR project reports in the literature. Here the search is for a completed EOR project with conditions similar to the one being considered for development.

In this chapter, application of EOR techniques is described as related to the initial phase of project selection. Overall reservoir screening is considered first. Next, reservoir evaluation methods used prior to the start of project development are discussed. Then, performance prediction is mentioned in terms of both laboratory studies and computer procedures. Finally, selected field histories are referenced as being representative of field projects completed with the different EOR processes and reported in the literature.

RESERVOIR SCREENING

Reservoir data are the basis for any reservoir screening procedure. Selection of an EOR process for a particular reservoir will depend on the data. Good and complete reservoir data will improve the chances of selecting the best EOR processing. Each developing country should make every effort possible to have available a data base that includes all possible information on each of its reservoirs. Much time and effort need to be expended by appropriate people in a developing country to ensure that information about each reservoir in the data base continues to increase. Steps need to be taken to obtain as much core and log data as possible whenever a new well is drilled. All geological and reservoir engineering parameters should be included. Reservoir rock and fluid properties are a most important part of the data base. Volumes of all kinds of fluids injected into and produced from a reservoir should be recorded.

Reservoir screening utilizes some kind of predetermined or in-house developed technical screening criteria. In screening, each reservoir in the data base is considered on an individual basis with regard to application of a particular EOR process. With screening, one or more of the EOR processes are indicated as technically feasible for a particular reservoir. Generally, screening will provide focus on the kinds of EOR processes that should be considered for further exploitation in a developing country. Reservoir size, with its indicated EOR process or processes, will play an important part in the selection program. It should be kept in mind that screening is only a first step toward final selection of a particular EOR process for a given reservoir. The final determination will depend on evaluation of the reservoir along with its predicted performance.

RESERVOIR EVALUATION

Reservoir evaluation is a most important step to ensure the success of an EOR process in the field. Time and money spent in defining the reservoir before development starts can help prevent costly failures. Basically, the evaluation needs to define as many of the reservoir's characteristics as possible. For example, the size of the box is important to obtain: (1) the EOR resource and potential reserve, and (2) the volume of injectants that might be required. The shape of the box tells much about the pattern type and spacing that will be required for development. Perhaps no single reservoir characteristic has led to more failures of EOR and other recovery projects as has reservoir heterogeneity. Under some conditions, if defined, reservoir heterogeneities could make the use of any EOR process impossible.

Different methods have been developed and are being used in reservoir evaluation. Here, geological and engineering, well testing, and tracer evaluation methods, three of the most significant techniques, are discussed. These methods are applicable in defining reservoirs in which any EOR process might take place. Every effort should be made in the developing countries to see that technology on how to use these methods is available. The three methods depend on data and results obtained from cores, logs, and field tests. Persons embarking on a reservoir evaluation program should anticipate certain related costs.

Reservoir or exploitation geological studies are key to any reservoir evaluation. Geologists trained in these specialities are a must if the study is to be successful. Unfortunately, geologists with a bent for exploration usually do not like to perform these studies. If a pilot project is being considered, log and core data need to be collected from the pilot area and surrounding wells in the field. For a full-scale field development, core and log data from all available wells should be used. A geological work sheet needs to be prepared for each well. The work sheet, among other things, should show porosity, absolute and relative permeability, fluid saturation, ion content of the in-place water, clay content and sensitivity, and so on. Correlations should be developed where core and log data are available from the same well. These correlations can be used to better define areas where only log data are available. Cross sections have to be prepared of the project area to show vertical heterogeneities. Fence diagrams are needed to indicate the anticipated areal conformance with the EOR process.

The July 1977 *Journal of Petroleum Technology* gives a good treatise on the role of reservoir geology. One of the articles describes how reservoir geology was used to characterize the layers of the Tensleep Reservoir in the Oregon Basin Field (5). Geological studies have been used to provide explanations for performance of different EOR projects. Cores were taken before and after in situ combustion (6) and surfactant flooding (7) projects, and geological studies were performed before and after to help in evaluating performance. If geology studies performed in a reservoir before a projet starts indicate problems, normally the project either will not be run or will be moved to another location.

A wide variety of engineering methods are used to evaluate a reservoir for an EOR process. One of the most important parameters to be determined is the oil remaining in the reservoir as the target for the EOR process. Another important parameter is the resistance to flow as oil and water flowing simultaneously through the reservoir. Different techniques have been reported useful in determining the residual oil present in a reservoir (8). Generally, more than one method is used since each procedure has both advantages and shortcomings. Knowledge of the resistance to the flow of oil and water is required to determine the mobility control conditions for an EOR process. Relative permeability data are required to make the mobility control calculation (9). The need for accurate and representative relative permeability data cannot be stressed enough.

Well test analysis is one of the easiest and lowest cost methods of reservoir evaluation. Pressure drawdown and buildup testing should take place in production wells throughout the field. Injectivity and falloff pressure testing should be used in the injection wells. Interference testing is of great value in determining reservoir continuity and in gaining insight into directional permeability problems. Steady-state pressure testing can be used to gain an idea of the relative permeability relationships that are acting in the reservoir. The U.S. petroleum literature is replete with information on how to perform tests and how to analyze the data. Technical monographs on the subject have been prepared by the Society of Petroleum Engineers (10,11). Technical organizations within developing countries who are either engaged in or anticipate becoming involved in EOR should acquire and study the available well testing literature. Several training courses are available in which the curriculum ranges from an introduction to the most advanced treatment of the subject.

Here, details are presented on injectivity testing to illustrate the value of well testing in reservoir evaluation. Test results obtained from existing wells prior to de-

velopment can be used to place project wells so as to take advantage of the determined directional permeability. Figure 3.1 shows the results from three sequential
injectivity tests run in a reservoir that was being considered for EOR after waterflooding (12). Water was injected at a given rate into an active injection well while
pressures were measured at this well and three or four observation wells. In two of
the three tests, injection and observation wells were interchanged during the test.
Injection and pressure data from these tests were used to determine the maximum
and minimum mobility values to water shown in the figure. As seen, the maximum
mobility to water generally runs northeast to southwest. To obtain the best areal
conformance with an EOR project in this reservoir, a line drive pattern would have
to be selected. Injection and producing wells would have to be placed in the north-
east-southwest direction.

Reservoirs being considered for EOR projects have been developed for at least
primary recovery operations and have production wells available. Other reservoirs
have undergone secondary recovery operations before being considered for EOR so
both injection and producing wells are available. Tracers can be injected into selected
wells in a reservoir, and the concentration and response time of tracers determined
in producing well fluids. These results provide keen insight into evaluating the reservoir in terms of possible vertical and area sweep problems.

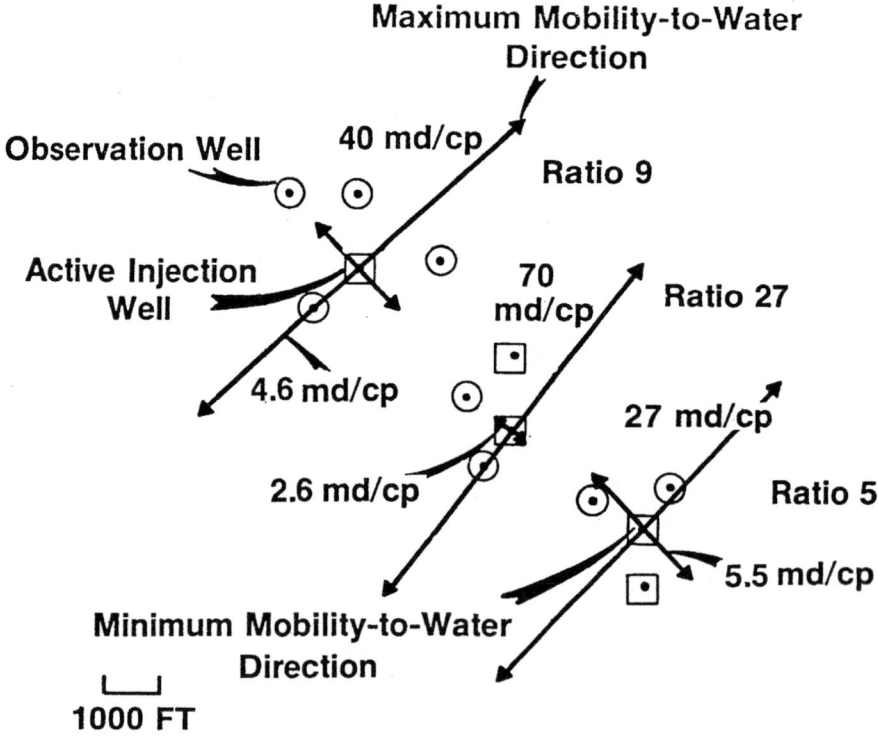

Figure 3.1 Interference testing. (*Source:* After Gogarty, Ref. 12)

Tracer tests can be conducted using either a gas or water injection well. For example, radioactive materials can be mixed with gas where predevelopment testing is being performed for a possible miscible or in situ combustion process. Tritiated water and thiocyannite are examples of tracers that can be used with water. In some cases, a concentration difference between the injection and in-place fluid can be used as a natural tracer. The difference between chloride concentration in injected and formation water is an example of a natural tracer.

Tracer studies cost time and money, but keep in mind that injected fluids normally represent the single biggest investment in an EOR process. Any loss of injectant efficiency because of unknown heterogeneities or other factors can have a disastrous effect on the EOR project performance and economics. The bottom line is that tracer studies are not only worthwhile from a technical standpoint, but help prevent colossal economic failures in the field. As indicated later in this section, tracer results also are valuable in keying a reservoir simulator to the field before predicting EOR process performance.

In the North Burbank surfactant flood project, tracer studies were used to determine the severity of channelling in the reservoir (13,14). Figure 3.2 shows logging results of one of the North Burbank wells (12). The gamma-ray log on the left defines the Burbank sand. The production log on the right gives an injection profile, which indicates that a severe vertical conformance problem exists within the reservoir. Thiocyannate tracer was injected to determine if an interwell channel existed between the injector and one of the offset producing wells. The results given in Figure 3.3, labeled as before treatment, confirm the interwell channel (12). As indicated in the figure, gelled polymer treatment was successful in reducing the permeability of the channel.

Figure 3.4 shows tracer results from the southwest part of the M-1 project (12). Tritiated water was injected into Well E-5 to obtain areal conformance information. Time lines are shown under some of the producing wells. The length of a time line indicates how long samples were taken from a given producing well. The magnitude of the radioactive tracer response at a given time is shown above the line. No tracer was detected in wells F-4 and F-6, even though these are two of the four wells surrounding injection well E-5. Significant and lengthy responses were observed in wells D-4 and D-6. Responses in B-4 and D-2 are significant since both of these wells are outside the inverter five-spot serviced by injection well E-5. These tracer results indicate the degree of heterogeneity that can exist in a reservoir. Application of an EOR process without understanding reservoir heterogeneity could lead to disastrous results.

PERFORMANCE PREDICTION

Both laboratory and computer studies are required in predicting the performance of an EOR process. The two methods attack the problem from different points of view. Both procedures are necessary, but neither gives the complete story. In this section,

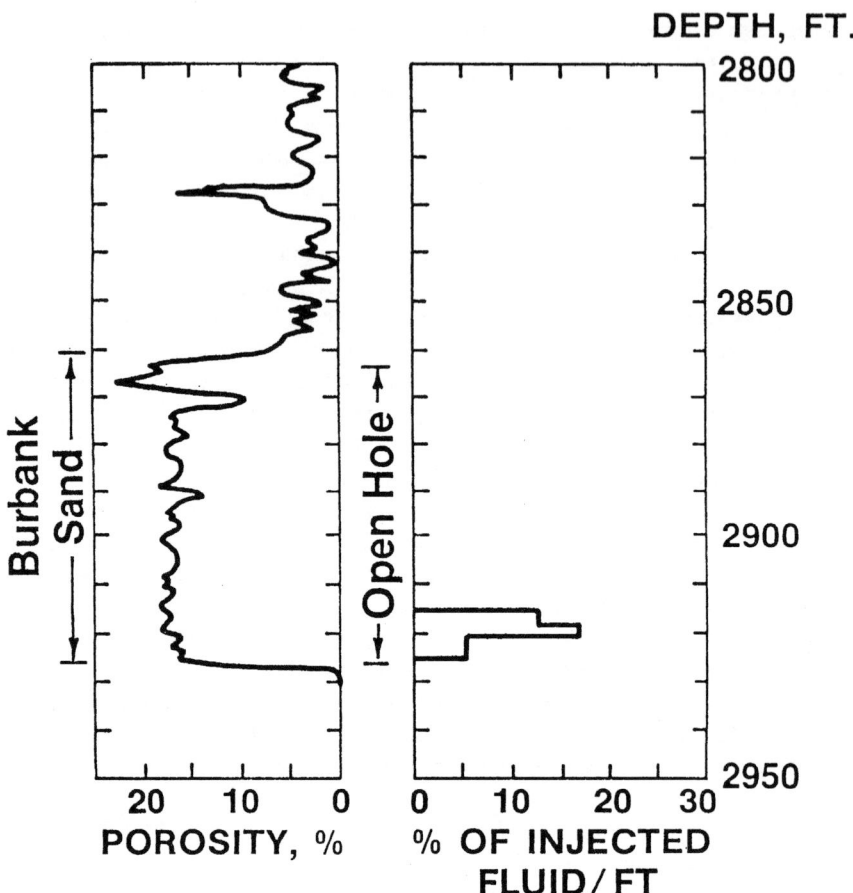

Figure 3.2 Profile of well 97-27. (*Source:* After Gogarty, Ref. 12)

methods for performing laboratory studies are considered first, then information is given on the use of various computer procedures to predict performance. It should be noted that both laboratory and computer results are used to decide whether a particular EOR process will be used in a given reservoir. Both are predictive tools. Even though the two procedures are discussed separately here, evaluation of results must always be made in concert. The real answer as to performance becomes available only after the EOR project is undertaken and is almost completed. Only then can the outcome be known with complete certainty.

A wide variety of laboratory procedures are used with the various EOR processes. Studies include those necessary (1) to design fluid systems for use in potential field projects, and (2) to obtain data what can be used to predict field performance. Measurements of rock and fluid properties and rock-fluid and fluid-fluid interactions are important in understanding the various EOR processes. Coreflooding is used to understand the physics of flow of a process in porous media and to gain some insight

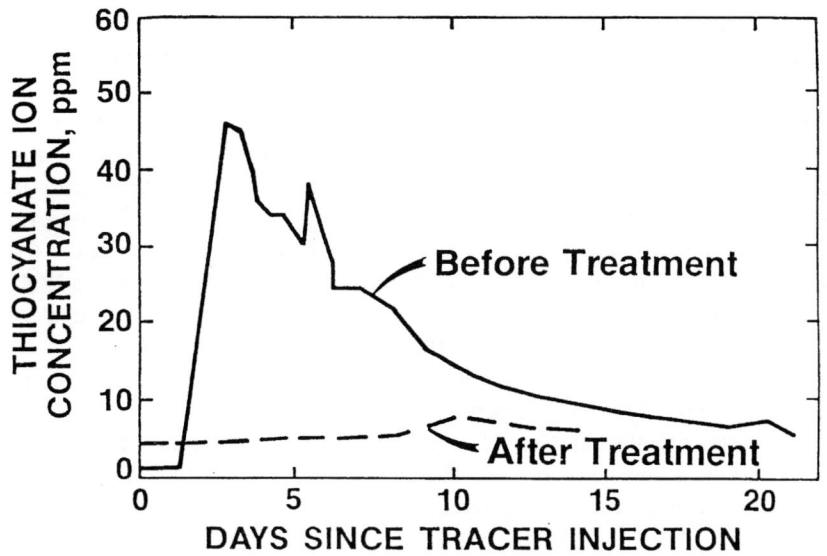

Figure 3.3 Tracer concentration for well 97-35. (*Source:* After Gogarty, Ref. 12)

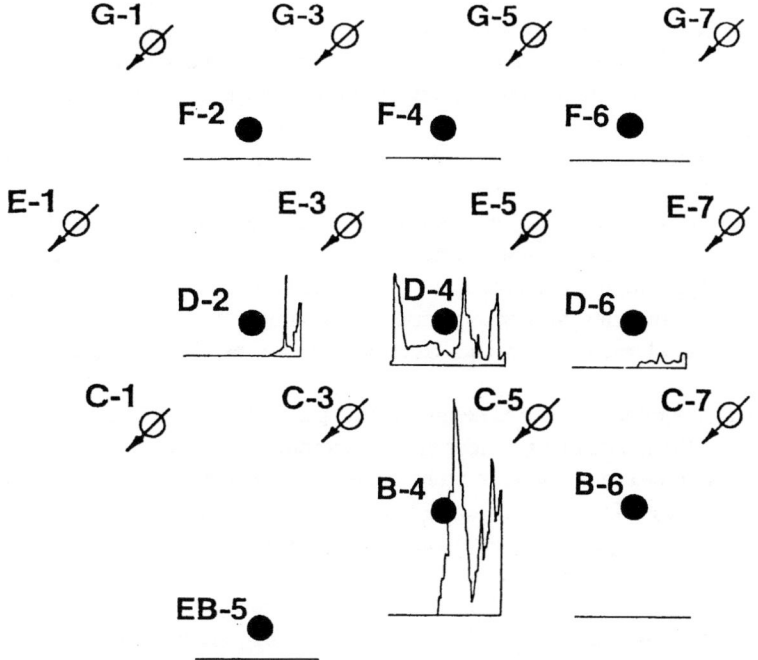

Figure 3.4 Tracer results from S-W part of M-1 project. (*Source:* After Gogarty, Ref. 12)

into performance efficiency. In coreflooding, every effort possible has to be made to ensure that (1) actual reservoir rock and fluids are used, (2) core materials are as near to virgin reservoir conditions as possible, and (3) cores and fluids remain at actual reservoir temperatures and pressures. Sophisticated and expensive equipment and highly skilled and technically trained personnel are required for these EOR laboratory studies. Developing countries will need to make a commitment to the laboratory phase if EOR work is going to be successful. Initially, the decision might be made to use contract services outside the developing country. In the long run, an in-country capability should be developed. A recent series of articles in the *Oil and Gas Journal* gives an indication of the equipment that is necessary to do EOR laboratory work (4,15,16,17). A brief discussion follows relating to the equipment that must be available for the three different categories of EOR.

For the chemical category, coreflooding and related equipment are very important. Here, pumps, core holders, pressure measuring devices, and so on are needed to perform linear, parallel, and radial corefloods. Radial corefloods are important in terms of polymer transport through the porous media (18). Linear corefloods are used in all aspects of chemical flooding, but are particularly useful in more fundamental surfactant and polymer flow studies (19,20). Mobility control is one of the most important aspects of chemical flooding (9,21). Relative permeabilities are required to determine the reciprocal relative mobility curves, and equipment is needed for these determinations. Analytical methods and associated equipment are required to make the measurements necessary to evaluate the chemical flooding processes. Among these are instruments to measure viscosity, interfacial tension, screen factors, chemical concentrations, and such. Surfactant polymer and alkaline agents are depleted as they move through the reservoir. Equipment is needed to determine these losses.

Miscible flooding normally takes place in the laboratory at high temperatures and pressures. Adequate coreflooding equipment must be available to operate in these conditions. Minimum miscibility pressure is a key parameter in miscible flooding. This pressure can be determined whether in a slim tube coreflooding apparatus or in the new type bubble equipment (22). PVT measurements play a most important part in the various miscible and immiscible flooding process. A PVT cell with its associated gas chromatograph and viscometer is a must for the miscible flooding program.

Thermal laboratory studies use equipment to perform steam floods and combustion experiments. Steam can be provided in the laboratory by either a small low pressure boiler or by an electrical induction tube heater. Steam is injected into one end of the core and produced from either the same or other end to simulate the cyclic or continuous process, respectively. Normally, the most important parameter to be determined is the steam/oil ratio. For combustion experiments, an adiabatic burn tube normally is used. Temperature control and heaters are positioned around the tube so that heat loss is essentially zero. For most experiments, the tube is packed with sand from the reservoir. The fuel requirement is perhaps the most important parameter determined from the combustion experiment. In both steam and combustion experiments, accurate fluid volumes must be determined as well as the composition of produced gases. A gas chromatograph is used for the latter purpose.

A wide range of computer procedures can be used to predict field performance of an EOR process. The range starts with a predictive or empirical model, which is keyed to available field data and ends with a sophisticated three-phase, three-dimensional finite difference reservoir simulator. Between these are the models that combine unit displacement values with vertical and areal sweep efficiencies.

Predictive models were used to calculate EOR performance of reservoirs in the 1984 NPC study (1). Table 3.1 provides information on the types of models used for each of the six EOR processes. These models were calibrated with data obtained from completed field projects. Essentially, these predictive models are curve fits of available field data. Where adequate field data from large projects were not available, reservoir simulators were used to project pilot project performance; then these results were used to calibrate the predictive model. Use of these models is now commercially available and could be used by technical personnel in developing countries for EOR performance prediction.

Recovery in any flooding process depends on the areal conformance, vertical conformance, and unit displacement factors. Different kinds of models can be used to estimate these factors; all factors are then multiplied together to obtain the recovery. Unit displacement for most EOR processes is taken directly from linear coreflood laboratory data. A Dykstra-Parsons type model is one of the ways used to estimate vertical conformance factors. A computer program can be obtained to handle these calculations. Variance factors are estimated from field core permeabilities and laboratory observed mobility ratios for the process. A streamline model is used to estimate areal conformance factors. In the simplest case, the streamline model assumes a homogeneous reservoir with displacing and displaced fluid having a unit mobility ratio. More advanced streamline models account for heterogeneous conditions by assigning different values of permeability to various regions of the reservoirs. This method of combining unit displacement with vertical and areal conformance factors is perhaps one of the easiest and lowest cost methods for developing countries to get into the business of predicting recoveries for EOR processes.

A reservoir simulator is the ultimate method presently being used to predict per-

Table 3.1
1984 NPC Study Predictive Models

Recovery processes	
Chemical	
polymer	–Stream tube model developed during NPC study
surfactant	–Modified DOE model
alkaline	–Modified surfactant model
Miscible	
carbon dioxide	–Extensively modified DOE model
Thermal	
Steam	–DOE model using modified gomaa algorithm
in situ combustion	–Model based on published correlation

formance of EOR processes. Use of this sophisticated predictive method requires a large amount of both laboratory and field data. Large computers and skilled personnel are required to run the models effectively. The models use three-dimensional finite difference grids and account for the different components located in the gas, the other two fluid phases, and on the rock. The need for a reservoir simulator must be considered in light of many factors, including cost and the acceptability of using simpler models. If a reservoir simulator is needed, a decision is required as to whether to buy, lease, or write. Writing a simulator is a long-term, high-cost, personnel-intensive program. If the decision is to buy or lease, simulators available from consulting firms should be considered since they are more field- and applications-oriented than those obtained from universities.

Simulators must be keyed properly with appropriate laboratory and field data before making predictions of full field performance. If the simulator is keyed with only laboratory data, a wide discrepancy may result between actual and predicted cumulative oil production. Simulation procedures for predicting EOR performance in the field include:

1. Matching laboratory waterflood and EOR flood results.
2. Matching field waterflood results.
3. Matching field tracer results, if available.
4. Fixing simulator parameters with laboratory and field matches.
5. Making predictive runs.

To illustrate these procedures, Figure 3.5 shows results of a simulator match of laboratory waterflood and polymer flood data.

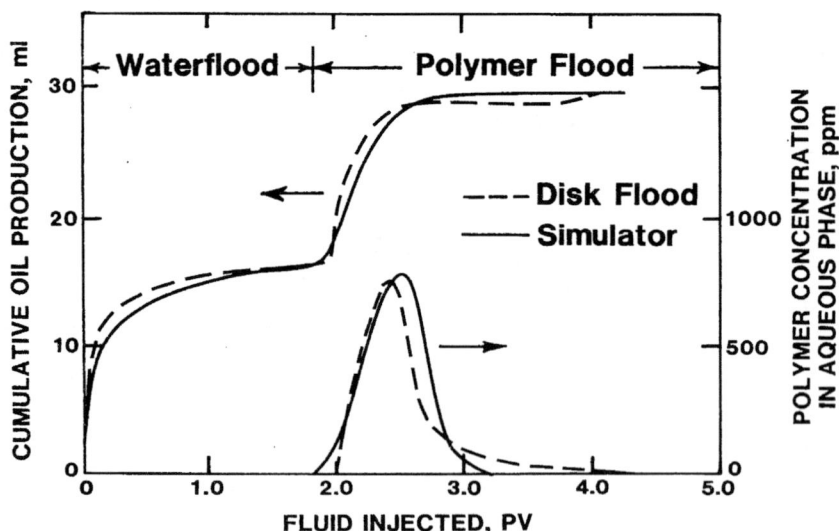

Figure 3.5 Simulator match of lab waterflood and polymer flood.

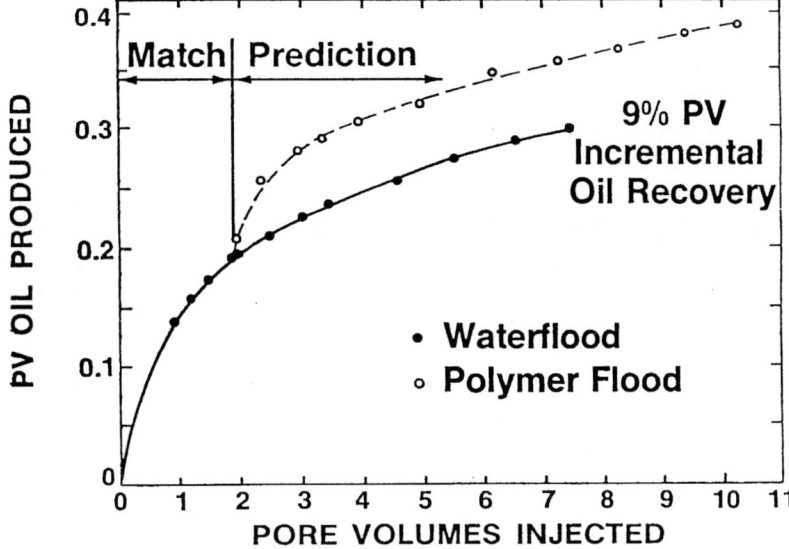

Figure 3.6 Simulator match for waterflood, and predictive runs for waterflood and polymer flood. (*Source:* After Gogarty, Ref. 12)

The laboratory match takes into account the various polymer flooding process variables, including rock-fluid and fluid-fluid interactions. As indicated in Figure 3.6, field waterflood results were matched to key the simulator to reservoir geology and heterogeneities (12). Tracer results were available and were used in the field matching process. After simulator parameters were fixed with the laboratory and field results, predictive runs were made for continued waterflooding and for the proposed polymer flood. The predictive results for these runs are given in Figure 3.6.

SELECTED FIELD HISTORIES

This section deals with selected field histories for the various EOR processes. Some representative histories are referenced for the EOR processes. The vast majority of published EOR field results in the world are found in U.S. literature. For this reason, representative histories were taken from that literature. The discussion of a particular EOR process includes some information about field projects in the developing countries and countries other than those in North America.

Polymer

Two methods of operation are being used in the application of the polymer flooding process. In one method, a larger pore volume (10–50%) of polymer solution is injected, and the project is operated as a polymer-augmented-waterflood. In the other method, a small volume of polymer solution (gelled on the surface or in situ) is injected to either reduced the permeability of high permeable zones or plug natural fractures in the reservoir, or both. Both polysaccharides and partially hydrolyzed polyacrylamide (PHPA) polymers have been used in the two methods of operation.

Examples of the polymer-augmented-waterflood method of operation are given by Marathon on their projects in the North Oregon Basin and Byron fields (23). The projects in these fields have led to a significant increase in oil production in both cases. Details are reported on the many aspects of conducting and evaluating a large-scale polymer-augmented-waterflood. A unique feature of these projects is that the injected PHPA polymer was manufactured at the field site. In this sense, these projects represent the most advanced technology in the world. The manufacturing technology is relatively simple and could be of great value for use in developing countries.

Many materials and techniques have been used to reduce water channelling in oil and gas reservoirs. A recent article by Esso Resources Canada, Ltd. reviews some of the literature on using polymer gels to control near-wellbore conformance problems (24). Results are presented on using phenoformaldehyde gels in the field because of their stability at higher temperatures. Stability is of primary concern when using polymer gels in high temperature reservoirs to improve vertical conformance in the near-wellbore region.

On a worldwide basis, the use of polymer flooding has been rather limited outside the United States. In the United States, numerous projects have been started. The Windfall Profit Tax (WPT) provided the incentive for the vast majority of these projects, especially those where gelled polymer was injected to improve vertical conformance problems. Outside the United States, the use of polymer for near-wellbore treatments has been rather limited, if not almost nonexistent. Two pilot polymer-augmented-waterfloods are being conducted by ONGC of India. Injection is underway in one of the projects and the wells are being completed for the other. Polymer floods have been conducted in USSR and the Federal Republic of Germany (FRG). Studies have been conducted by laboratories in the U.K. to determine the temperature stability of polymers for use in the North Sea. Polymer flooding is being considered by China for use in some of its fields.

Surfactant

The history of surfactant flooding reveals the development of two kinds of fluid systems. One system used a large pore volume (up to 50%) of a low concentration (less than 2.5%) of active surfactant. The other used a small pore volume (5 to 15%) of a high concentration of surfactant. In some cases, polymer was included within the surfactant system for mobility control. In both systems, polymer solution was used for mobility control behind the surfactant solutions. Early reported laboratory work distinguishes between the two kinds of systems (25,26,27,28,29).

The beginning of modern-day surfactant flooding in the field was reported by Marathon (30). Results are given for both a three-quarter and 40-acre project. Performance of these two early projects was truly pheonomenal and created interest throughout the world in surfactant flooding. Marathon, which has used both gas oil and crude oil base petroleum sulfonates as the surfactants in its projects, has continued to develop its technology with these surfactants and has conducted numerous projects both alone and jointly with other countries. Exxon has moved from petroleum sulfonate to synthetic surfactants. It reported successful results with a synthetic surfactant in their small (0.68 acre) London pilot test (31). Larger scale projects are now being run by Exxon to confirm small pilot results.

Surfactant flooding is perhaps the most technically complex and costly of the EOR processes to apply. Use outside the United States has been limited. Two surfactant flooding projects have been completed in France. A pilot project is being designed for use in the Anklesvar field of India. China is moving toward the use of surfactant flooding in their large Daqing oil field. The Russians have reported using surfactants in some of their fields. They have probably been the first to try (unsuccessfully) nonionic surfactants in the field.

One surfactant field project has been tried in the Federal Republic of Germany. Research is underway in Norway and the United Kingdom to develop surfactant systems (sans polymer) for use in the North Sea. Surfactant flooding is underway in the Handil field of Indonesia. Argentina is considering a pilot surfactant flood using technology from the United States.

Alkaline

Of the three chemical methods, alkaline flooding appears to be the most controversial. Some people feel that the process has great potential. Others believe that successful field results are not possible. The controversy centers on consumption or loss of the alkaline agent through reaction with the reservoir rock and fluids. Basically, the process develops surfactants in situ when the alkaline agent reacts with the natural acids in the crude oil. Unfortunately, the reactivity of the alkaline with reservoir rock and fluids may be as great, if not greater, than with the crude. Loss of the alkaline agent causes the process to revert back to the waterflood with no interfacial tension lowering.

Both successful and unsuccessful alkaline field projects have been reported. One of the recently reported successful projects is the alkaline flood in the Isenhour unit, which is located in Wyoming (32). This project is unique in that soda ash (Na_2CO_3) from nearby mines is being used as the alkaline agent. Perhaps the high concentration of this low-cost chemical, along with the high pore volume of polymer used for mobility control behind the alkaline slug, is the reason for the apparent success of this project. At least one company is betting that alkaline flooding is the wave of the future for chemical and other EOR methods. Shell has reported that it will be starting an alkaline flood in the White Castle field with its improved technology (33). Essentially, the procedure includes mixing a surfactant with alkaline solution before injection.

Alkaline field projects outside the United States appear limited. Wells are being drilled by ONGC for a pilot alkaline flood in India. Russia has reported on a limited amount of field work. Ammonium hydroxide was used as the alkaline agent for a field trial in Hungary.

Miscible (Immiscible)

Miscible gas injection with methane started in the United States in the early 1950s. At present, carbon dioxide is being used because miscibility is achieved at a lower pressure. In higher temperature reservoirs, nitrogen sometimes is injected rather than carbon dioxide because of better miscibility conditions. Most of the large carbon dioxide floods in the United States are taking place in the large carbonate reservoirs of West Texas. Natural carbon dioxide is being used for these floods and the gas is being supplied by pipeline from fields in Colorado and New Mexico.

The granddaddy of the carbon dioxide projects is in the SACROC unit (34,35). Both large-scale secondary projects and a small-scale tertiary project have been conducted in this unit. Published results on miscible flooding give an indication of its potential and show the types of problems that might be encountered. General knowledge of the overall aspects of miscible flooding probably is best obtained from Monograph No. 8 published by the Society of Petroleum Engineers (36). In addition to this document, much information on miscible flooding from the standpoint of field design and application is available in the literature associated with Gulf's Little Knife CO2 Minitest (37).

As mentioned earlier, miscibility may not be possible because certain reservoirs are too shallow to permit high enough pressures. Field projects have been undertaken in some shallow reservoirs using carbon dioxide under immiscible conditions. In a test in the Wilmington field, carbon dioxide was injected for the purpose of swelling and reducing the viscosity of the heavy oil in one of the reservoirs in that field (38). Details on how these two factors effect recovery are discussed in the reported results. Perhaps the largest immiscible carbon dioxide flood in the world is being conducted in Marathon's giant Yates field in West Texas. Here, the in-place oil viscosity is not high in terms of heavy oil standards. Nevertheless, the operator apparently feels the swelling of the oil and the reducing of its viscosity would be sufficient to increase recovery.

Miscible flooding is a process that shows promise for use in lower permeability-heterogeneous reservoirs with and without natural fractures. Even though reservoirs of this type are found in the developing countries throughout the world, activity there has been minimal. Canada has been a leading country in using the hydrocarbon miscibility method. There are over 25 projects underway in Canada. One immiscible carbon dioxide project is being conducted in Italy. Libya has conducted a hydrocarbon miscible project in its Intisar "D" field. Carbon dioxide is being used in Hungary for EOR. In Algeria a hydrocarbon miscible project is ongoing in the super giant Hassi Messaoud field. China is considering the use of carbon dioxide miscible projects in some of its fields.

Steam

Of the various EOR methods, steam injection is the most advanced and is being applied more frequently throughout the world. Both cyclic and continuous steam injections are being used in these projects. In some projects, hot water is injected rather than steam, which condenses into hot water after being injected. Generation of steam injectant is relatively simple from a technological point of view. In most projects, steam is generated in low pressure boilers and produced crude is used as the fuel. Cogeneration steam projects are also underway. Here, high pressure steam is generated and used first to provide electricity. Low pressure steam from the electrical generators is then injected into the reservoir for steamflooding. With the exception of water treatment and corrosion problems, steamflooding technology is not too complicated in shallow reservoirs. Cyclic steam can be started in one or more wells in a field as a kind of a pilot program. Continuous steam injection on a large scale then can be applied if the cyclic results warrant this action. As reservoirs get deeper, insulated tubing and/or downhole steam generation must be considered. This technology is much more complicated and costly.

The vast majority of field projects in the United States are being conducted in California. Activity has decreased somewhat with the decrease in crude prices. Steamflooding technology is rather mature, and all aspects of thermal recovery, including this technology, are disucssed in the SPE Monograph No. 7 (39). Because so many steamflooding projects have been reported in the literature, papers giving state-of-the-art reviews appear more appropriate here than the discussion of a particular steamflood field project (40,41). These papers give field examples and discuss the factors that are important for a technical understanding of steamflooding.

Steamflooding has been widely applied in countries outside North America. It is being routinely used in the heavy oil reservoirs of Venezuela. These projects are one of the best examples in the world of EOR technology being successfully applied in a developing country. Projects also are underway in Trinidad and Tobago and more are being planned. Also, projects are being planned in the Federal Republic of Germany and in Italy. Work is underway on steamflooding projects in the USSR, China, and India. Romania has considered the use of steamflooding.

In Situ Combustion

In situ combustion projects use high-pressure, high-capacity compressors to handle the injected air. Availability of this kind of equipment, along with skilled personnel to operate and maintain the equipment, may limit in situ combustion use in developing countries. Special considerations are needed in well completions. As the combustion front reaches a production well, temperatures increase rapidly. Casing, tubing, and pumping equipment need to be designed to account for these conditions. Gas pressures at injection and producing wells may be or may become high and blowout preventive measures are required. The bottom line is that careful consideration must be given to designing and operating a field combustion project.

As with steamflooding, much literature is available giving design requirements and presenting the results of successful field projects. The thermal recovery monograph applies to in situ combustion as well as steam projects (39). Many review articles have been completed dealing with successful field projects. The state-of-the-art review of field projects by Chu probably is more valuable than selecting a report on one specific project (42). This article reviews 25 successful and nine aborted projects. All of the design factors necessary to develop a successful project are given in this article. Outside of the United States and Canada, in situ combusion field projects are limited. Pilot projects are being developed in India and Hungary. In situ combustion work is underway in China relating to field development of its heavy oil reservoirs.

4

EOR State-of-the-Art

This chapter deals with topics that relate to or have an influence on the state-of-the-art of EOR. First, a brief status report is given on various EOR categories. Next, selected technical factors are considered as related to the application of EOR methods in the field. Then economic considerations show how economic calculations are made for the purpose of project evaluation. Finally, various factors that influence EOR development are discussed.

STATUS OF EOR TECHNIQUES

EOR activity is considered first by using the number of projects both in the United States and in different parts of the world. Next, U.S. production figures are presented as a means of comparing the development stage of processes in the various EOR categories, followed by comments that relate to further technical advances required to provide emphasis for increased EOR application. Last, a brief review presents ongoing activities in some developing and other countries.

Every two years the *Oil and Gas Journal* publishes a survey of EOR activity in the United States (43–46). Table 4.1 shows the results of this survey in terms of the three EOR categories for the years 1980 through 1986. Survey results are published in the spring and are representative of conditions at the end of the previous year. As seen in the table, the chemical category had the most active projects in 1986. This result is somewhat misleading in that polymer flooding accounts for over 85% of the projects in the chemical category. In fact, the number of surfactant and alkaline projects has remained essentially constant during the time of these surveys. Based on the results given for chemical, polymer flooding technology appears to be approaching the proven development stage. Again referring to Table 4.1, the increase in miscible (immiscible) projects was due largely to an increase in carbon dioxide miscible and immiscible activity. A slight increase also has taken place in using hydrocarbon and nitrogen gas for miscible displacement. Most of the miscible (immiscible) projects started since 1984 are in large reservoirs and represent application

Table 4.1
Active U.S. EOR Projects

	1980	1982	1984	1986
Chemical polymer surfactant alkaline	42	85	138	206
Miscible (immiscible) carbon dioxide hydrocarbon other gas (nitrogen, flue gas)	34	50	84	105
Thermal steam in situ combustion hot water	150	139	151	201
Total	226	274	373	512

on an anticipated commercial basis. Only time will tell if this supposition is correct from both a technological and economic basis. All the increase in thermal projects shown in Table 4.1 is the result of steamflooding. Hot water in 1986 replaced three of the otherwise essentially constant number of in situ combustion projects. Steamflooding appears to continue to dominate the thermal category. Where heavy oil reservoirs are available, steamfloods are being used apparently with little risk of technological failure.

It should be noted that the results in Table 4.1 are representative of the higher U.S. crude prices that existed through most of 1985. Nevertheless, the relative number of the kinds of projects in any one of the three categories will probably remain fairly constant, thus still indicating the same order of technology development.

Table 4.2 compares EOR activity outside the United States for 1984 and 1986 (45, 46). Both ongoing and planned projects are indicated. The majority of the projects are from Canada, Venezuela, and the FRG. Other countries represented include Trinidad and Tobago, the Congo, Indonesia, France, Italy, the U.K., Tunisia, Hungary, and Holland. Note that projects are not included for the USSR, China, India, or the Middle East even though EOR activities are known to be underway in those countries.

In Table 4.2, projects in the chemical category have more than doubled, but the activity remains at a low level. The increase is due primarily to additional polymer projects. The limited increase in polymer activity is surprising in view of the high level of polymer technology available in the United States. As indicated in the table, miscible (immiscible) activity outside the United States is about constant, if not decreasing. Most of the projects are using hydrocarbon gas and the majority are taking place in Canada. The use of carbon dioxide may be somewhat unique to the United States because of its large natural sources of that gas. Thermal projects are seen in Table 4.2 to have more than doubled, with the overall level almost half of that in

Table 4.2
Ongoing or Planned EOR Activity Outside U.S.

	1984	1986
Chemical	5	12
polymer		
surfactant		
alkaline		
Miscible (immiscible)	32	24
carbon dioxide		
hydrocarbon		
other gas		
(nitrogen flue gas)		
Thermal	39	81
steam		
in situ combustion		
hot water		
Total	76	117

the United States. As in the United States, the majority of the thermal projects are using steam, with projects in Venezuela and Canada accounting for the largest share. Several hot water floods are taking place in the FRG. The high activity in steam (hot water) is a good indication that this technology is at a stage where use in developing countries might be possible.

U.S. EOR production for 1982 through 1986 is shown in Table 4.3 (1, 44–46). For 1982 NPC results are compared with those of the *Oil and Gas Journal* survey. The data in this table give an indication of the potential of some of the EOR processes and their stage of development.

Consider the production data for the chemical processes. Most of production here

Table 4.3
U.S. EOR Production (barrels per day)

EOR processes	1982		1984	1986
	NPC	O & GJ	O & GJ	O & GJ
Chemical	4,410	4,409	13,398	16,901
polymer				
surfactant				
alkaline				
Miscible (immiscible)	50,660	71,915	83,011	108,216
carbon dioxide				
hydrocarbon				
other gas				
(nitrogen flue gas)				
Thermal	450,200	298,624	364,560	479,669
steam				
in situ combustion				
hot water				
Total	505,270	374,948	460,969	604,786

is from polymer projects (see Table 4.1). This increasing production indicates that polymer technology is being applied successfully and is in an advanced stage of development, as compared to the other chemical methods. The relatively low level of production is indicative of polymer flooding potential. Here, many of the polymer projects are recovering only small volumes of incremental oil by correcting near-wellbore conformance problems. Even in full-scale polymer-augmented waterflooding projects, recovery of incremental oil is low because no mechanism is present to reduce the interfacial tension with its corresponding decrease in residual oil saturation. Herein lies the promise of surfactant and alkaline flooding. With its lowering of interfacial tension and mobility control, the recovery potential is as high, if not the highest, of any of the EOR processes. Unfortunately, the stage of development of the surfactant and alkaline processes limits their use to selected reservoirs in areas where adequate chemicals and technical personnel are available.

The miscible (immiscible) process data in Table 4.3 are somewhat of an enigma. The difference in the 1982 production data is the result of the NPC study reporting only on carbon dioxide miscible projects, whereas the *Oil and Gas Journal* survey covered all gas injection projects. Nevertheless, the 1986 miscible figure might be anticipated to be higher. Many of the projects from which the production is coming are in large reservoirs and by this time should have experienced significant increases. Certainly, these projects were started with anticipation of their great potential. Probably, the best that can be said at this time about the stage of the miscible (immiscible) process development is that a "wait and see" attitude will need to be taken. As with alkaline and surfactant flooding, miscible flooding can and is being applied successfully. The question still needs to be answered about how wide an application is possible.

The data for thermal in Table 4.3 is mostly representative of steam projects. As seen, the increasing level of producing rates indicates that steam has great potential, is a viable process, and is in an advanced stage of technological development. In the 1982 column, the NPC results are higher than the *Oil and Gas Journal* because data from more projects became available through the diligent efforts of the study group. It is unknown whether similar differences will exist in later years.

Further technological advances are required if the chemical processes are to be more widely applied in the field. Polymer is a key element in all the chemical processes. Improvements are required in all aspects of stability, including temperature, mechanical degradation, chemical, and so on. Temperature and chemical stability of surfactants also must be improved. Development of more cost-effective chemical systems is a major need. Quantity and types of chemicals need to be considered in this program. Injectivity and propagation of all chemicals must be improved, including an increased understanding of the many aspects of fluid-fluid and fluid-rock interactions. The apparent breakthrough in alkaline flooding technology needs to be pursued wherein a surfactant is added to the injected solution so as to increase the concentration of alkaline agent allowed before reaching the minimum interfacial tension point. Methods of predicting chemical flooding performance need to be improved by developing better reservoir simulators.

Technological improvements are needed to improve effectiveness in the field of processes in the miscible category. Better procedures are required to establish the minimum miscibility pressure of a gas that is being injected into a reservoir. Along these lines, improved methods of studying phase behavior are required relating to process efficiency. Reservoir simulation is an area of miscible flooding where additional work is needed. Methods of obtaining better mobility control are required because of the wide difference between the viscosity of the injected gas and the in-place oil. Improvements are needed in both gas processing and production technology, especially where carbon dioxide is used as the injectant.

Even though thermal is the most advanced of the EOR technologies, further improvements would increase recovery efficiency in the field. Gas foam technology needs to be improved so as to increase recovery efficiency by reducing gravity over-riding of the steam. Consideration should be given to injecting noncondensible gas with steam as a method of increasing recovery. Work needs to be continued on using thermal recovery in light oil reservoirs. Better hydraulic fracture treatments need to be developed so as to more effectively produce the heavy oil. Improvements are needed in insulated tubing and downhole steam generation to extend the use of steam to deeper reservoirs. Additional field trials are needed to increase the technology of using steam in naturally fractured reservoirs.

Research facilities and development laboratories engaged in EOR work provide an indication of EOR status and interest throughout the world. Many private and university centers are engaged in EOR work in the United States, Canada, and countries in Western Europe such as the U.K., France, the FRG, Holland, Norway, and Italy. In eastern Europe, the USSR is active in all aspects of EOR research and development. Other eastern European countries with laboratory programs include Hungary, Rumania, and Turkey. In South America, the Intevep facility in Caracas, Venezuela, is engaged in many EOR projects. In the Middle East, Bahrain is involved in an extensive program to study various EOR processes. The United Arab Emirates is working on EOR activity. Saudi Arabia, Tunisia, and Kuwait are looking into using the miscible flooding process. ONGC has an active EOR laboratory program at its Institute of Reservoir Studies at Ahmedabad. EOR research and laboratory programs are underway in China, both in Beijing and in the Daqing oil-producing area. The Japanese Oil Company is interested in EOR, but real research interest in Japan is from the standpoint of producing better chemicals such as surfactants and polymers.

TECHNICAL FACTORS

Conducting successful EOR projects in the field depends on a wide variety of technical factors. Some of these factors are process-specific and others are process-independent. Volumes could be written about the factors in both the process-dependent and process-independent categories. Here, examples are given to illustrate some of the different kinds of technology needed for successful EOR application in the field.

These examples are by no means all-inclusive, but they are presented to make readers aware of the kinds of problems that must be considered.

The examples in this section include those relating to pilot testing, project development, and project operations and evaluation. Many of the examples come from projects in the chemical category. Factors in the miscible and thermal categories, although not in most cases identical, will be similar. For those in developing countries, the important lesson here is to be aware of the vast amount of field-related technology required for successful EOR application. Generally, field experience is needed to learn the lessons. One way of circumventing costly mistakes is to be taught by people with field experience. This type of training is discussed later.

Pilot Testing

Pilot projects are run to obtain recoveries that can be used to predict full-field results. Usually, the game plan is to reduce expenses by keeping to a minimum the amount of injectant, the number of wells, the spacing between wells, and so on. With this plan, an isolated five-spot pattern is often used, or in some cases, an inverted pattern with one injection well surrounded by four producing wells. In other cases, a regular pattern configuration is used where four injection wells surround a producing well.

In an inverted pattern, all of the injectant displaces oil from within the pattern area. This is the good news. The bad news is that all of the oil being produced may not come from within the pilot area. It is therefore difficult to determine accurately the actual amount of oil produced resulting from the total amount of injectant used. This ratio of oil produced to injectant used is an important EOR parameter. Efficiency and costs are associated with this parameter. When the pilot ratio is subject to considerable uncertainty, the result does not mean much when extrapolated to a full-field development.

With some inverted pilot projects, the number of producing wells is reduced to cut expenses further. In the limit, only an injection well is used as was the case for the Little Knife project (37). In these cases, the ratio of oil displaced to injectant used is determined, and this ratio is subject to considerable uncertainty. Further, the need to change the displaced ratio to one produced in some kind of pattern causes additional uncertainty.

The situation in the regular five-spot pattern is the opposite of the inverted pattern. Here, all of the produced oil is normally from within the confines of the pattern. But the amount of injectant entering the pattern from the four injectors is difficult, if not impossible, to determine. Again, the ratio of oil produced to the amount of injectant used is not obtained with any degree of certainty. For both the regular and inverted five-spot cases, the ratio is even more difficult to determine where the area has not been previously waterflooded before undertaking the EOR pilot.

A question of importance is whether to use isolated versus repeated patterns to obtain pilot flood recoveries for predicting full-field performance. Table 4.4 compares the results of the Henry pilot and 119-R field projects in which essentially the same surfactant fluid system was used in the same reservoir. Note that 63% of the

Table 4.4
Pilot and Field Recoveries

Pilot/field project	Size (acres)	Tertiary oil recovery (%)
Henry	0.75	63
119-R	40.00	38

oil in place was recovered in the 0.75-acre pattern of the Henry test (7). In the 40-acre 119-R project, the recovery was only 38% of the oil in place (47). These results indicate the need for large projects with repeated patterns to determine full-scale recoveries.

Project Development

Project development requires both production and reservoir engineering skills. Usually, the tasks used in EOR operations are similar to those required for primary and secondary operations. However, the complexity of the tasks requires a higher level of engineering know-how and the availability of more adequate technology. The construction of surface facilities, completion of wells, and the selection of pattern types and spacings are given here as illustrative of the need for a high level of engineering. The placement of wells in the reservoir to achieve the maximum areal conformance and the use of injectivity testing to ensure proper system performance in the field are given as examples of advanced technology.

Surface facility design varies widely for the various EOR processes. Selection of correct construction material and the need for preventing corrosion are common requirements for all EOR processes. The surfactant flooding process is used here to illustrate the kind of production engineering that goes into providing the surface facilities. Other EOR processes will be similar. Work starts with site preparation, including associated roads, pipeyards, miscellaneous buildings, and electrification. A production system must be constructed that will separate oil and water containing significant amounts of surfactant, polymer, and other chemicals. To say the least, this is far beyond the type of treating system used in water flood operations. The injection system treats, filters, mixes, or manufactures the chemicals, and injects the different fluids used in the surfactant flooding system. In the water processing system, water is filtered and softened. The slug system provides for heating, filtering, and mixing the various components. The polymer system provides similar services, but in some cases a field polymer manufacturing unit will replace the mixing equipment. All tanks and lines need to be coated to prevent either corrosion or interaction with the various chemicals. An elaborate control system is required to regulate the flow of fluids in both the production and injection system. Quality control instruments are placed throughout the system to provide information to ensure that operations are to specification. With the high cost of the fluids being handled, mistakes can be expensive. Pumps and control methods must be selected to handle special problems such as shear degradation of polymer.

   Well drilling and completion are of primary importance in EOR operations. Care must be exercised to ensure that the formation in the near-wellbore region does not cause problems with the EOR injectants or produced fluids. This statement also applies to the stimulation and completion methods used for wells. Usually, greater care must be taken for injection as compared to production wells. For example, wells into which polymer will be injected may be cased through the formation and perforated. Here, the number of perforations per foot is very important. Too few perforations will cause high shear rates as injection takes place and the polymer will be shear degraded. Consideration needs to be given to the type of material used in the casing material across the formation and in the tubing. In some projects, fiberglass casing may be used across the formation so that induction logging can be used to determine saturation changes. Tubing materials may range from coated mild to high alloy steel. Temperature, corrosion, and fluid interaction will dictate the best type of material to use. In some steam projects, economics would indicate that the tubing be insulated.

   EOR projects in older reservoirs required careful scrutiny of the conditions of all wells previously drilled in the project area. Communication in old wellbores between different zones or reservoirs in the field can lead to real problems in an EOR project. Loss of expensive injectants by this means is not obvious at the surface, and these problems can cause real frustration in trying to understand project failures. An effective abandonment and reabandonment program is required to prevent these kinds of problems.

   Pattern type and spacing (48), including size, can have a profound effect on EOR performance. Figure 4.1 illustrates how pattern types affected performance of a particular surfactant flooding process. Laboratory disk floods for the 119-R and 219-R projects with the same fluid systems and volumes as used in the field recovered about the same amount of oil. Both projects were conducted in the same reservoir with

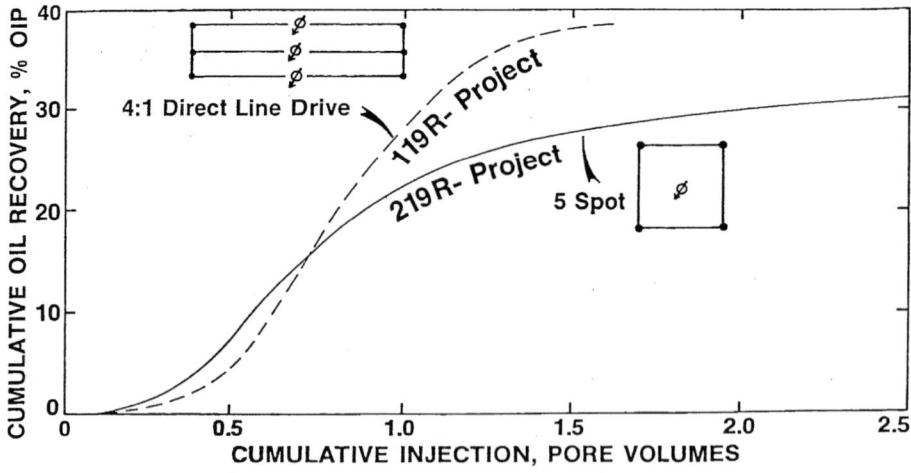

Figure 4.1 Effect of pattern type on oil recovery. (*Source:* After Gogarty, Ref. 48)

about the same spacings. The difference in field recoveries appeared to have been caused by pattern types. Results in Figure 4.1 indicate that line-drive patterns are superior to five-spot patterns for this process. In developing any field, well locations, injectivity, and pattern type should be given careful consideration.

Well locations for an EOR project are extremely important in obtaining maximum recovery. The goal is to attain the best possible areal conformance in the project area. Details of a highly successful surfactant flood are presented here as proof of the well location thesis (12). Figure 4.2 shows a map of the HXa sand, upper ter-

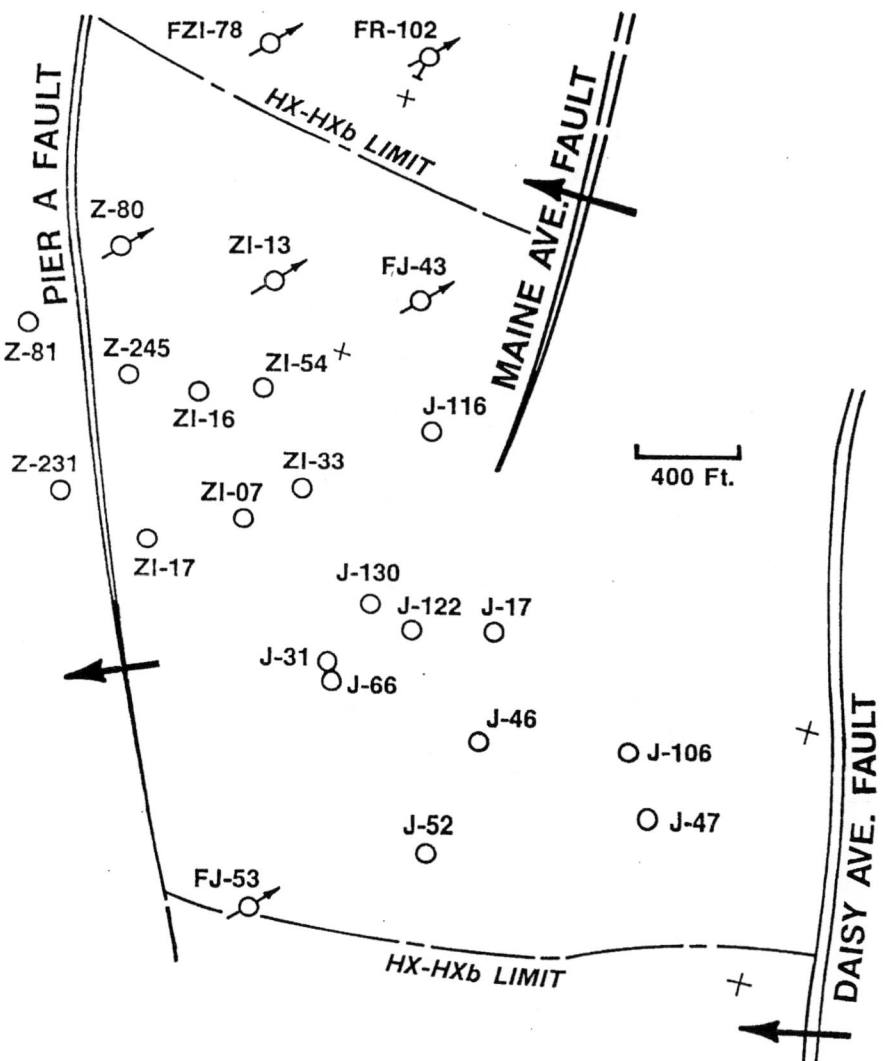

Figure 4.2 HXa sand, upper terminal zone—Wilmington field. (*Source:* After Gogarty, Ref. 12)

minal zone, Wilmington field in California (12). Before starting the U.S. DOE surfactant project in this field, well locations had to be selected. Interference testing indicated that the Pier A fault was sealing. Streamlines and corresponding flood fronts were calculated with a computer model.

Results are shown in Figure 4.3 for the first well locations selected (12). Seven existing and three new wells were used for this proposed layout. Injection and producing rates are shown for each well used in the study. The calculated streamlines and flood fronts indicate the areal conformance. Gaps between the streamlines of the various sets of injection and producing wells indicate a poor areal conformance.

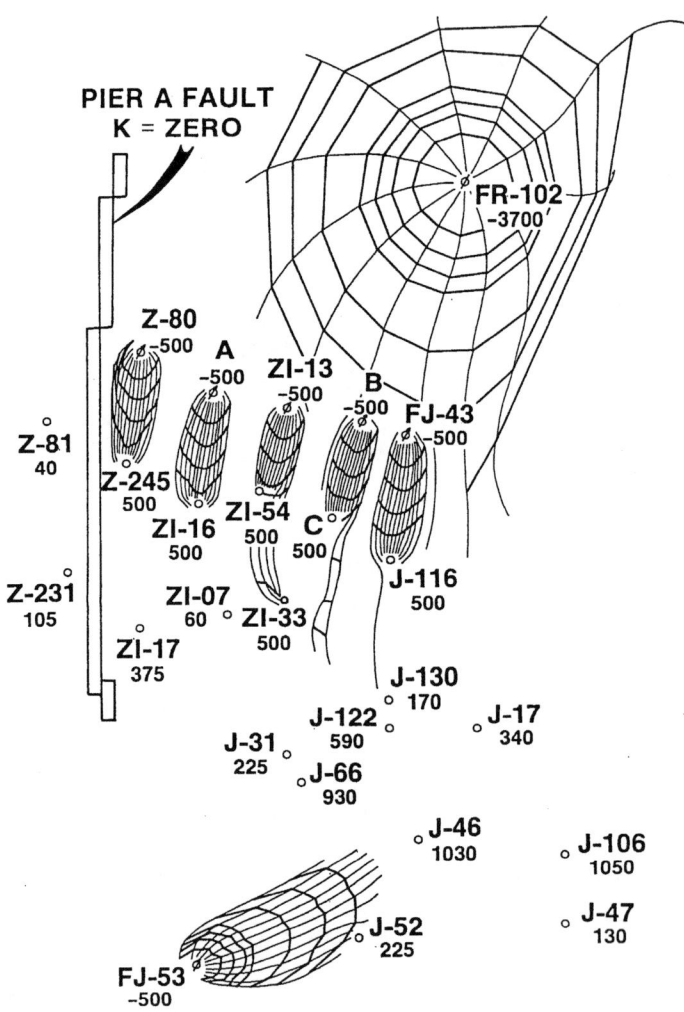

Figure 4.3 Initial proposed pattern—Wilmington DOE micellar/polymer project. (*Source:* After Gogarty, Ref. 12)

Generally, water flow caused by the high pressure gradient from north to the south caused this poor conformance.

Figure 4.4 shows the pattern selected for the Wilmington project (12). Of the nine wells used, five were new and four existed previously. Wells were located so that the sealing Pier A fault backed up the injection wells. With these well locations, the computer results indicate good areal conformance. The project has been completed successfully since these calculations were made. Based on different assumptions, ultimate oil recovery was 35 to 52% of the oil in place at the end of water flooding. A value of 43% appears reasonable. One reason for the success of this project may be the high areal conformance resulting from the well locations.

Injectivity testing is important prior to the start of injection in an EOR project. These tests are used to determine injection rates in the field and to see if any skin damage or other plugging problems are caused by the EOR injectants. Usually a

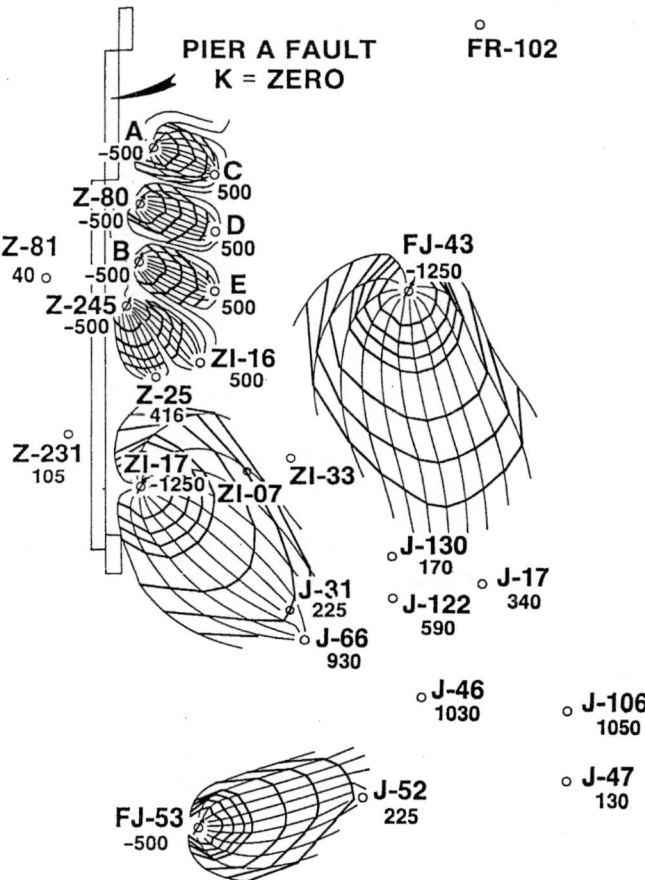

Figure 4.4 Final proposed pattern—Wilmington DOE micellar/polymer project. (*Source:* After Gogarty, Ref. 12)

nonproject well is used in the testing program. If this is not possible, a project well can be used before the project starts or early in its life. The problem with using a project well is one of timing. The other project wells may have been drilled before the injectivity test is completed. If the injectivity test indicates that the project must be aborted, the wells and associated costs are lost.

A polymer flood project is used here to demonstrate injectivity test principles. Test results are given in Table 4.5. This injectivity test and others in the Tensleep formation were conducted to obtain field data for a fieldwide polymer flood (23, 48). The average flow rate and the pressure drop between the sandface and reservoir were used to calculate the injectivities. These injectivities were normalized with the value for water obtained before polymer injection. Normalized injectivities decreased with polymer injection as would be expected when the higher viscosity polymer solution is injected. If the normalized injectivities had remained at 1 as polymer solution was being injected, the results would have indicated that polymer was being sheared while passing through the perforations into formation. With water injection, the normalized injectivities increased. These results indicate that polymer injection did not plug the wellbore.

Project Operations and Evaluation

Because of the high costs associated with EOR projects, ongoing operations should be monitored closely and on a continuous basis. Early detection of problems often can prevent costly and catastrophic failures. The best engineering talent available should be used to evaluate an EOR project's performance throughout its life. In no case should a project be completed and then evaluated. Production history with its associated fluid analysis, the use of observation wells, and injection/production logging methods are discussed here as representative of ways to evaluate project operations.

Plots of oil production with time are an effective way of evaluating project performance. These plots should be made in terms of oil rate and oil cut. Plots are helpful for both the total project and individual wells. The pore volume of oil produced should be followed for any confined pattern within the project as well as the total project. Monitoring of produced fluids by well testing and measurements of

Table 4.5
Polymer Injectivity Field Test Data

Average (rate) (B/D)	Average polymer concentration (ppm)	Stabilized wellhead pressure (psi)	Normalized injectivity
3,220	0	140	1.00
3,280	250	170	0.98
3,270	520	220	0.93
3,250	1,080	320	0.84
3,030	0	180	0.90

produced water can be used to evaluate the project. Checks should be made to see that the summation of fluids from individual well tests closely parallels production measured at the production battery. Material balance calculations should be made as a check to ensure that all fluids are moving properly into and out of the project reservoir. Pressure histories of injection wells should be studied continuously as an indication of project performance.

An analysis of produced fluids provides valuable information about EOR project performance. The arrival times of different fluid banks give an indication of how mobility control is operating in the process. In chemical EOR for example, sulfonate analysis indicates the arrival of the surfactant fluid bank. When polymer is mixed in fresh water, chloride analysis provides information about the arrival of the polymer bank. Chemical or radioactive tracers can be injected at different stages of an EOR process. Among other things, data from these tracers will give information about the volumetric sweepout efficiency and direction of fluid movement. Water analysis for inorganic ions is helpful in determining problems related to water quality control, corrosion, and pollution.

Observation wells are extremely valuable in providing data to assist in evaluating ongoing EOR project operations. When reservoirs are shallow in depth, core tests may be drilled and evaluated both during and after project operation, and the results of these core tests will supplement the information obtained from the observation wells. With deeper reservoirs, core test costs become more prohibitive, and observation wells become one of the best means of obtaining continuous data from within the reservoir.

Usually, observation wells are positioned between the injection and production wells. Locations normally are selected to be representative of average reservoir conditions within the pilot or project area. So-called active and dry completion observation wells are two of the types used. With an active completion, casing is set through the formation and perforated. Fluid samples can be obtained repeatedly from the reservoir with this kind of well. Also, steady-state pressure testing can take place by using the values obtained in the observation well with those determined in the offset injection well. With the dry completion, casing is set through the formation in such a way as to prevent any reservoir fluid from entering the wellbore. The observation well is then logged repeatedly during ongoing production operations. These logging operations can provide important information about changes in preflood and postflood oil saturation, arrival time of different fluid banks, size of oil bank, and vertical sweep efficiencies. To perform correctly, the logging tool must be able to measure formation properties behind the casing. Both carbon-oxygen and induction logging have been used for this purpose. High-resistivity fiberglass is placed across the formation when induction logging is to be used.

Injection/production logging is used as a method of evaluating fluid movement into and out of a reservoir. Frequent logging runs should be made throughout the life of an EOR project. Runs should always be made in an injection well before and after any injection fluid changes are made, for example, polymer solution to water or vice versa.

Correct injection/production logging tools need to be selected for the particular EOR process being studied. Exxon's Louden project is an example of the care that must be taken in selecting the correct tools (49). In this project, an extensive study was conducted on measuring injection well profiles in viscous solutions. Results indicated that injection/production logging tools developed to measure waterinjection profiles would not work satisfactorily while injecting viscous surfactant and polymer solutions, and further the water injection tools had to be modified to perform satisfactorily. Since that time, extensive tool modifications have been made and tested. The swing arm tracer (SWAT) tool is not the choice for logging in viscous fluids (50). Remember that service company tools must be operating correctly when determining injection profiles in any EOR process.

Economic Considerations

In the final analysis, economics determines whether full-scale EOR projects will be conducted. Economic calculations start with a prediction of crude oil production from the EOR process. Next, oil production is converted into a revenue stream by using projected crude oil prices. Processes' independent and dependent costs are deducted from the revenue. In some cases, inflation factors are applied to oil prices and process costs. Finally, different kinds of economic indicators are applied to determine project feasibility. These economic concepts are discussed in this section.

Because of the economy of scale, economics of full-scale EOR projects generally will be better than those of pilot operations. Developing countries may choose to run so-called uneconomical pilot projects to obtain valuable technical information. These projects are very important from the standpoint of developing technology even when the economics are far from favorable. The key for a given country is to proceed in an orderly manner with laboratory and field research and development of applicable EOR processes. This strategy should include pilot testing that will ensure that adequate technology is available when economics are favorable for field-wide application. This kind of strategy is being used in India and China, among other places. There, several pilot projects either are underway or are being planned. It should be kept in mind that even though pilot projects may not be profitable from an economic point of view, recovery and cost data serve as invaluable starting points for predicting economics of full-scale development.

Anticipated crude oil production serves as a basis for economic analysis. Oil production as a function of time must be estimated for the entire project life. Techniques for making these predictions vary widely and will depend on the technical resources available in a country. Any of the procedures described in the performance prediction section of this book could be used to make these estimates. Predictive models of the type given in Table 3.1 may be of particular value to developing countries.

There are certain economic risks versus rewards that must be considered when evaluating which EOR process should be used in a given reservoir. The selection of one of the three chemical processes for a particular reservoir is used to illustrate these risk versus reward conditions. Parameters for this reservoir are given in Table

Table 4.6
Reservoir Parameters Used For Calculating Oil Recovery

Project area	1,000 acres
Pattern type	Inverted 9 spot
Pattern spacing	40 acres
Net sand thickness	90 feet
Porosity	17%
Oil saturation	37.5%
Depth	5,000 feet
Permeability	75 md
Crude oil gravity	25° API
Crude oil viscosity (reservoir)	15 cp
Oil cut	10%

4.6. Figure 4.5 shows the estimated cumulative oil production by different recovery methods (12). At the start of the EOR project, 39.3 million bbl of oil remained in the reservoir. Continued waterflooding was estimated to recover an additional 11.8 million bbl of oil. With polymer flooding, 14.6 million bbl of oil were projected to be recovered. The increase represents 10% of the oil remaining in place after waterflooding. Surfactant flooding was projected to recover 24 million bbl of oil. The increase in recovery equates to 45% more of the remaining oil in the reservoir after waterflooding.

Potential for EOR recovery (reward) with surfactant flooding is higher than with polymer flooding. Recovery with alkaline flooding probably falls between the other two chemical methods. Whereas rewards are higher, risks are higher with the high

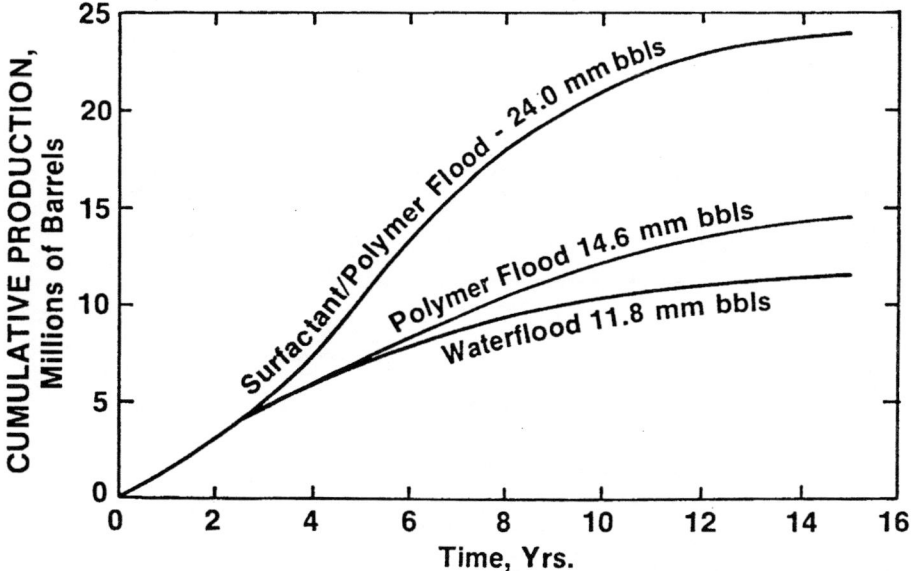

Figure 4.5 Cumulative oil production by different recovery methods. (*Source:* After Gogarty, Ref. 12)

front-end investment projects using surfactant flooding. Risk associated with project failure can be restricted with polymer flooding, but rewards are lower. Economic parameters, such as discounted-cash-flow rate of return, favor the low front-end investment polymer projects, even though oil recovery is lower.

Forecasting of crude oil prices is at best a speculative task. Figure 4.6 is from the 1984 NPC study and presents a historical average of U.S. crude oil prices up to 1983 (1). Prices continued to drop well in 1986, reaching a low value of about $10 per barrel and are presently at about $18 per barrel. The wide fluctuation in crude oil price starting in 1973 lends credence to the statement that "forecasting oil prices is black art."

Many economic evaluations are performed with a fixed oil price. Inflation factors are not used in these fixed oil price studies. Usually, studies include economic calculations at more than one oil price. Process-independent and dependent costs are a function of energy cost; hence the price of oil. Fixed oil price studies normally relate to fixed costs for a given price per barrel of oil. As the price of oil is changed in a study, the process costs either are increased or decreased accordingly. Usually, inflation factors are used with process costs only when they are used with oil prices.

Process-independent costs are for those expenses that are the same for any EOR process. Table 4.7 gives examples of onshore process-independent costs. The cost for direct annual operations for secondary recovery is needed in those cases where EOR costs are being considered as incremental above those associated with primary and secondary recovery. For example, many of the chemical and miscible flooding processes will be applied in reservoirs that have been previously waterflooded.

Table 4.8 shows the types of process-dependent costs related to the three major EOR categories. Anyone doing economic calculations will need to spend consider-

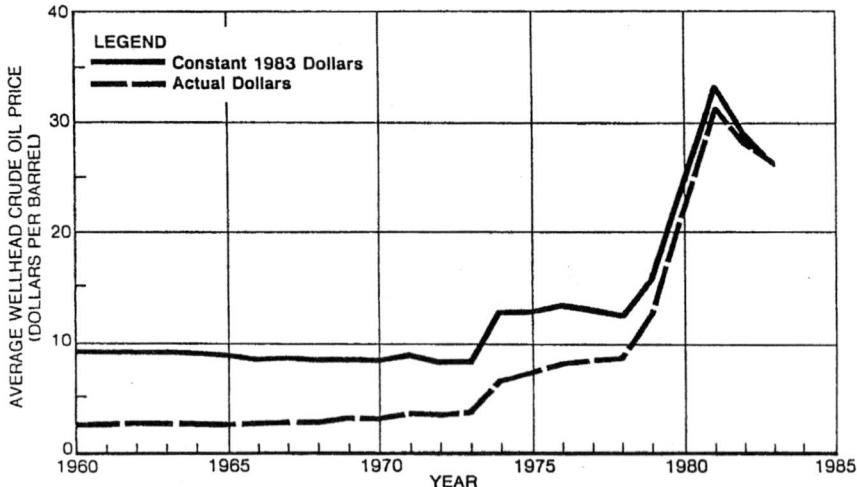

Figure 4.6 Historical average U.S. crude oil price trends. (*Source: Basic Petroleum Book, Petroleum Industry Statistics,* Vol. 3, No. 3. American Petroleum Institute, Washington, DC, November 1983)

Table 4.7

Onshore Process Independent Costs

Direct annual operations for secondary recovery
Converting a producing to an injection well
Drilling and completing production and injection wells
Equipment for a new producing well

able time and effort in predicting the costs associated with the particular process being evaluated. Items listed in Table 4.8 under chemical flooding pertain to polymer, surfactant, and alkaline flooding. Under miscible flooding, the cost information stresses carbon dioxide. For thermal recovery, costs are presented both for steamflooding and in situ combustion.

Usually, the time value of money is considered to determine if EOR processes are economically viable. Cumulative cash flow versus time relationships are developed by using oil production with its corresponding price and the related process cost. Taxes may or may not be used when generating these cash flow relationships. Discounted cash flow rates of return as determined from these data are used for project evaluation. The net present value at a given discount rate and undiscounted net profit also are used for evaluation. Economics ratios are sometimes used, including the net present value to investment, net profit to investment, and investment efficiency. Investment efficiency is defined as the ratio of the total discounted cash flow to the maximum cumulative negative discounted cash flow (1). The investment efficiency technique appears to have value for ranking the heavily front-end-loaded EOR projects. These projects not only have high front-end capital investments, but also have exceptionally high operating costs.

Table 4.8

Process Dependent Costs

Chemical Flooding
    chemical costs
    chemical formulations
    chemical injection plant cost
    well workover cost
    operating expenses
Miscible Flooding
    carbon dioxide supply cost
    carbon dioxide injection plant investment
    investments for well and surface equipment
    operating expenses
Thermal Recovery
    steam costs
    air costs
    injection and production well costs
    cost of surface facilities
    produced water recycle plant cost
    cost of waste gas treatment
    operating expenses

Factors Influencing EOR Development

Many factors influence EOR development. Some factors can be affected by the governments of the developing countries involved. Other factors relate to the stage of oil development in a country. Still other factors depend on the availability of existing infrastructure and natural resources.

In this section, recovery potential and oil price are considered first as perhaps being two of the most important factors. Next, the effect of losing surface equipment and wells from existing reservoirs is considered. Then, the role of government is discussed from the standpoint of taking advantage of EOR and of stimulated development. Finally, the effect on development of in-country supplies of EOR injectants is considered in terms of manufacturing capability and natural resources.

Recovery potential and oil price are two of the most important factors in providing incentive for EOR development. The recovery potential depends on (1) the oil saturation in the reservoir at the start of EOR operations, and (2) the recovery efficiency of the EOR process being applied. For example, most reservoirs being considered for thermal recovery have high initial oil saturations because only a small amount of the heavy oil present has been recovered by secondary operations such as waterflooding. The recovery efficiency of the various EOR processes depends on their state of technological development, as discussed earlier.

Oil price is independent of technology and, in some respects, is related to worldwide geopolitical conditions. Generally, EOR projects required large front-end investments and long payout times. Oil prices must be high and remain either constant or increase during the project life if the EOR development is to be economically attractive. Consider Figure 4.6 where oil prices within the United States are seen to be subjected to wide swings (1). Also, remember that EOR process costs are energy (oil-price) dependent. Assume that EOR projects were initiated in the United States in the early 1970s when oil prices (energy and development costs) were low. If most of the production from these projects occurred when oil prices were high in the early 1980s, then the projects could have been economic bonanzas. With EOR projects started in 1984, the opposite price trends were taking place and the economic results could have been disastrous. A developing country can spur EOR development by guaranteeing high and constant oil prices. This approach was used in the United States to provide incentive to Union Oil for developing its oil shale process in Western Colorado. Here the U.S. government guaranteed an oil price of $38/barrel for oil produced by the process.

As reservoirs become depleted by primary and secondary operations, consideration should be given to plugging the wells and salvaging the surface equipment. Loss of wells and surface facilities probably represents lost opportunities for EOR development in certain reservoirs. Every effort needs to be made in developing countries to consider EOR projects before these resources are lost. The added cost of redrilling wells and replacing production facilities well cause EOR projects to be even less attractive from an economic standpoint. In offshore operations, replacement costs of platforms for drilling and operations will prohibit, in most cases, if not all, any EOR development.

EOR development can benefit a developing country in many ways. Increases in a country's domestic oil production will reduce foreign oil import requirements and their corresponding payments. EOR technology will make it possible to recover a higher percentage of oil from existing and new reservoirs. With higher recovery efficiencies, a nation can expand its recoverable resource base. Conducting EOR projects will lead to the development of more skilled personnel in the country. Ancillary industries will need to be developed to provide materials for the different EOR processes, for example, the surfactants and polymers used in the chemical processes. EOR provides a viable alternative to the development of other energy sources. Each country should consider EOR before looking at other resources such as coal liquids, tar sands, and oil shale. Taxes and royalty payments from EOR development can be important in funding other government services.

Government incentive programs in developing countries can play an important role in stimulating EOR development. Different methods were used in the United States starting in 1974. These techniques are mentioned here as examples of what a government might do. From 1974 through 1980, the U.S. Energy Research and Development Administration and the DOE cost-sharing program provided direct financial support (51). DOE's Enhanced Oil Recovery Incentives Program from 1970 to 1981 allow front-end recoupment of expenses by letting companies sell controlled oil at higher prices (52). The Crude Oil WPT Act of 1980 reduced tax rates for EOR projects (53). The Economic Recovery Tax Act of 1981 provided research and experimentation tax credit (54).

Table 4.9 shows the effect of incentive programs on distribution of 1982 active U.S. chemical EOR projects. The projects in the 1982 *Oil and Gas Journal* survey (44) are distributed in the four categories shown in the table. The only effect of the WPT incentive program has been a large increase in polymer flooding projects. This increase took place because the WPT act favored low risk, low front-end investment projects.

EOR development can be stimulated to some extent by the availability of injectants in a country. For example, the capability of being able to manufacture surfactants and polymers may lead to the development of EOR processes that use these chemicals. This kind of stimulation might be enhanced if manufacturing facilities were sitting idle. The cost of chemicals might be considered somewhat differently if plant costs had been made and personnel were not working. On the other hand, the absence in a country of any manufacturing facilities for chemicals will certainly be a deterrent to this kind of EOR development.

Natural resources in a country also can be important in stimulating EOR devel-

Table 4.9

Effect of Incentive Programs on Distribution of 1982 Active U.S. Chemical EOR Projects

	Preincentive	Cost sharing	Front end recoupment	WPT (March 1981)	Total
Polymer	16	1	11	20	48
Surfactant	7	7	6	4	24
Alkaline	4	2	5	2	13

opment. Economics of flooding with carbon dioxide depend on adequate low-cost supplies of this chemical. Countries with reservoirs containing carbon dioxide will want to consider developing EOR projects with this material. Carbon dioxide from reservoirs is probably the most economic source. But, separation from off gases of fertilizer and electrical generating facilities should not be overlooked. Many developing countries, such as India and China, have the potential of using gases from these kinds of plants. Natural deposits of inorganic salts also may serve as a stimulus for EOR development. For example, soda ash from natural deposits might be useful in alkaline flooding and stimulate its development.

5

EOR Training

This chapter provides information on EOR training that is available to developing countries. First, technology transfer methods are discussed in terms of different kinds of courses. Comments are presented on course selection and evaluation and some of the advantages and disadvantages of the various methods of providing training or technology transfer. Second, the different kinds of training programs are described as related to the types of organizations used as the training facility. Here, advantages and disadvantages are also mentioned and some examples are given relating to the kinds of organizations that provide training programs or facilities.

TECHNOLOGY TRANSFER

A wide variety of techniques can be used to provide technology transfer of EOR information to a developing country. The scope of any course should be determined by those who will be receiving the instruction. Methods of presentation, course outlines, and instructors' qualifications need to be reviewed and approved by appropriate people in the developing country. A decision must be made whether to bring instructors to the developing country or to send students to locations where courses are offered.

An axiom of any technology transfer is "there are courses and then there are courses." In other words, some are good and some are bad. Also, good courses could have bad instructors and bad courses could have good instruction. Neither of these conditions is good, but the latter may be the lesser of the two evils. Course and instructor evaluations are the keys to obtaining effective EOR technology transfer for a developing country. Before a new course is used in a country, information about the course and its instructors should be obtained from every possible nonprejudiced source. As different courses are presented both in and outside the country, thorough written evaluation should be made by students of both the instructors and courses. These evaluations should be collected, compiled, and kept at a central location. Evaluation information should be used to select future courses and instructors.

Many general and selected topic EOR courses are listed as covering more or less the information presented earlier in Part I. Unfortunately, the general course may be too general and the specific courses too specific. The trick is to select a course that will be of most benefit to the students who will attend. For example, general courses usually give an overview of all the EOR processes and related information. This kind of course may be of value to management for informational purposes, but technical personnel will not receive the training necessary to perform specific EOR tasks. Numerous selected topic courses may be required to cover the various aspects of even one of the EOR processes. In this case, a training outline is required to ensure that enough courses are provided to cover all of the details of the EOR process selected. One of the porblems with a sequence of courses may be that each course starts with a general overview. These overviews can unnecessarily lengthen individual courses and waste valuable student time for the entire program.

Generally, private in-house courses are more cost efficient. In these courses, the entire spectrum of a particular EOR process can be covered in as much depth as required. Course outlines can be developed to meet the specific needs of a particular developing country. The level of information presented can be tailored to fit the specific needs of those selected to attend a given course. Further, example problems and reservoir conditions used in the course can be taken from situations that exist in the country. Students from the developing country may travel outside their country as a group to where an in-house course is being presented. However, economics normally will dictate that the instructor or instructors travel to the country of choice.

A variation on the theme of a private in-house course is where instruction alternates between theoretical and practical aspects. Here theoretical concepts are applied immediately in the field. The course starts in a developing country during the early stage of an EOR project. The instructor or instructors live, teach and work in the developing country with the technical people who are responsible for the project. First, theoretical principles are discussed in the classroom. Then these same principles are applied on a practical basis in the field. In this instruction, course examples and problems are used directly and immediately either in the laboratory, on the computer, or in the field. Here, students and management know almost immediately the worth of the instructors and course. A real advantage of these kinds of courses is the close personal contact between instructor and students. Usually the ratio of students to instructors is small, and many different groups of students may interact with the same instructor or instructors on many different aspects of the same project.

EOR courses generally are based on the knowledge of an individual or group of individuals who have been associated with research and development of EOR processes. These courses, of necessity, must include information from the published literature. In evaluating a course, care should be exercised to see that the content is up-to-date in terms of (1) the basic knowledge of the people offering the course, and (2) the literature and techniques being used to present the course. A significant amount of literature exists in a large number of professional journals throughout the world. Additional information is presented at technical meetings and published each year. Government reports contain valuable information on both completed and ongoing

EOR projects. The bottom line is to include in the evaluation of any EOR course a determination of the extent of its included published literature. In fact, a part of any EOR training might well include a course or courses based on a thorough search of the published literature.

The wave of the future in EOR instruction is in the use of (1) personal computers to handle EOR data bases and make appropriate calculations, and (2) video cassettes as a means of providing the most up-to-date information. Both of these techniques should be considered when evaluating EOR courses. Personal computers and TV videos should be looked upon as resource supplements. These techniques will seldom, if ever, be an end in themselves.

TRAINING PROGRAMS

Different types of organizations are considered here as related to providing EOR training programs. Organizations furnishing these programs include universities, oil and gas companies, consulting companies, industrial training companies, and consultants from both universities and the private sector. These different organizations provide a wide spectrum of EOR courses and programs. Some organizations focus on laboratory related courses. Others present information more related to the reservoir engineering aspects of EOR and may or may not rely heavily on the computer. Still other organizations specialize in production operations and field-related problems. A most difficult task in selecting an organization for EOR training is to be sure that courses are broad enough to cover the total subject, but not so broad that necessary details are lost.

Table 5.1 lists types of organizations that have EOR training capabilities and gives some examples of companies with in each of these organizational types. Some of the larger consulting companies and the industrial training companies have offices at several different locations throughout the world.

Under the universities category, all of the example schools are in the United States. Other universities within the United States and schools in Canada and Western Europe also provide EOR training programs. A university course may utilize as instructors professors from one or more departments. For this reason, many of the courses are presented on the university campus. The university courses are long on laboratory and theoretical work, up-to-date on the published literature, and short on field experience and applications. The reason for this shortcoming is that the number of professors with extensive industrial experience is limited.

The oil and gas companies shown is the table are typical of those that can be a valuable resource for EOR training. Many of the larger companies have their own in-house training programs. These kinds of companies have access to large amounts of actual laboratory and field EOR data. Both EOR reservoir and production engineering skills are practiced at a high level of competence in most companies. High-speed computers with related programs are available and used in the companies to analyze and evaluate their EOR projects. Developing countries need to learn how to

Table 5.1
Organizations With EOR Training Capabilities

Universities
Penn State University
Stanford University
Texas A&M University
University of Texas
Oil and gas companies
British Petroleum
Conoco
Exxon
Marathon
Shell
Consulting companies
Core Laboratories
Improved Petroleum Recovery Int., Ltd.
Scientific Software—Intercomp
Surtek, Inc.
Industrial training companies
International Human Resources Development Corporation
Oil & Gas Consultants International, Inc.
H. K. van Poollen and Associates, Inc.
Consultants (private-university)
Applied Reservoir Technology, Inc.
W. Barney Gogarty & Associates, Inc.
S. M. Farong Ali
S. A. Holditch and Associates, Inc.
Mungar Petroleum Consultants, Ltd.

tap the vast EOR knowledge resource within these companies. The key may lie in trading exploration and production concessions for EOR technology and training programs.

The difference between the example consulting and industrial training companies shown in Table 5.1 is only a matter of degree. EOR training by a consulting company may be somewhat of a secondary business. For this reason, their training programs must be evaluated carefully to see that they are up-to-date. Usually the industrial training companies' most important line of business is their EOR and other training programs. Most of the people working in these training programs, as well as those in the consulting companies, have come from industry. Many of these people are experts in their field and can and do bring much knowledge to the training programs. The key to evaluating these programs and their instructors is to see if they are up-to-date.

Under the last organizational type in Table 5.1, examples are given of consultants who are operating small-scale businesses. Normally one and probably no more than two professionals operate out of these kinds of businesses. The list could include hundreds of people. Almost every petroleum engineering professor throughout the world can and will act as a consultant. Many qualified people have left the industry either through early retirement or other means. Some have years of valuable EOR experience, and some have become private consultants. The key here is for the developing countries to match the available talent to their EOR training needs. An example of where this matching procedure has been used effectively is through the UNDP in India and the other developing nations.

6

Conclusions and Recommendations

1. Potential reserves from EOR recovery are high in developing countries. Oil is in known reservoirs and represents a national resource base. Look at EOR as a method of recovering an otherwise unrecoverable national resource.

2. Six EOR processes in the chemical, miscible, and thermal categories are being developed worldwide. To a greater or lesser degree, all processes are based on reducing interfacial tension and improving mobility control. Focus research and development activities along the lines of improving these process variables.

3. Screening criteria provide information on the stage of EOR development. Published criteria are applicable on a worldwide basis. Use the 1984 NPC screening criteria for reservoirs in developing countries.

4. EOR injectants available in a country can act as a stimulus for development of a particular process.

5. Use complete geological and engineering studies in the EOR project selection phase. Time and money spent in the laboratory and field for this phase provides insurance against project failures.

6. Develop and continually update a data base on existing and new reservoirs within a country. Screen this data base as a means of directing EOR development.

7. Leave no stone unturned in evaluating a reservoir for possible EOR application. Well testing and tracer techniques are particularly effective in defining the reservoir.

8. Use laboratory and computer studies to estimate EOR process performance. Understand that neither technique gives the complete story, but both procedures are necessary. Reservoir simulators are the ultimate tool for use in computer studies. Simulator development and operating costs are high.

9. Design methods, operation procedures, and evaluation techniques are well documented for each of the six EOR processes. Use literature as a means of moving technology into a developing country.

10. Worldwide activity indicates that steamflooding is the most advanced EOR process. Consider using steam injection as a recovery method for any heavy oil reservoir.

11. Potential for recovery is the highest with alkaline and surfactant flooding of

any of the EOR processes. Further technological advances are required before these processes can be used widely in the field. Applications of polymer flooding, though at a high stage of development, are taking place mostly in the United States.

12. Miscible (immiscible) activity outside of the United States is about constant. Carbon dioxide is being used for a large number of projects in the United States. Outside the United States, hydrocarbon gas is being used.

13. Understand that large amounts of laboratory and field technology are required for successful EOR applications. Some technology is process dependent; other technology is process independent. Both laboratory and field experience are needed by those who are providing EOR technology to developing countries.

14. Use repeated versus isolated patterns for pilot testing where possible.

15. Know that the complexity of EOR projects requires a high level of engineering know-how and the availability of adequately trained personnel. Exercise care in designing and constructing surface facilities. Give well completions, abandonments, and reabandonments special consideration.

16. Never wait until an EOR project is completed before making an evaluation. Use every method possible for evaluation, including analysis of injection and production data, produced fluid analyses, observation wells, and injection/production logging.

17. Use economic analysis to determine whether full-scale EOR projects should be run. Remember so-called uneconomical pilot projects provide important technical information. Pilot testing now ensures the availability of adequate technology when economics for full-scale application are favorable.

18. Consider using different types of government incentives as a means of stimulating EOR development. Use price supports for EOR oil; participate in the cost of EOR pilot and larger scale projects; and provide tax reductions for EOR projects.

19. Course outlines and instructors' qualifications are keys to successful technology transfer to a developing country. Some courses are good and some bad. Good courses can have bad instructors and bad courses can have good instructors. Set up an evaluation system to ensure that all EOR training is of the highest quality.

20. Be sure that the organization furnishing EOR training is broad enough to cover the total subject, but not so broad as to miss the important details. Organizations are not as important as the experience of instructors who are providing the instruction.

References to Part I

1. National Petroleum Council. 1984. *Enhanced Oil Recovery.*

2. Office of Technology Assessment, Congress of the United States. 1978. *Enhanced Oil Recovery Potential in the United States.*

3. Taber, J. J., and T. D. Martin. 1983. "Technical Screening Guides for Enhanced Recovery of Oil," paper SPE 12069 presented at the 1983 SPE Annual Technical Conference and Exhibition, San Francisco, October 5–8.

4. Goodlett, G. O., M. M. Honarpour, H. B. Carroll, and P. S. Sarathi. 1986. "Lab Evaluation Requires Appropriate Techniques—Part 1," *Oil & Gas J.,* pp. 47–54.

5. Morgan, J. T., F. S. Cordiner, and A. R. Livingston. 1977. "Tensleep Reservoir Study, Oregon Basin Field, Wyoming-Reservoir Characteristics," *J. Pet. Tech.*, pp. 886–902.

6. Hewitt, C. H., and J. T. Morgan. 1965. "The Fry In Situ Combustion Test—Reservoir Characteristics," *J. Pet. Tech.*, pp. 337–342.

7. Gogarty, W. B., and H. Surkalo. 1972. "A Field Test of Micellar Solution Flooding," *J. Pet. Tech.*, pp. 1161–1169.

8. Cordiner, F. S., D. T. Gordon, and J. R. Jargon. 1972. "Determination of Residual Oil Saturation After Waterflooding," paper SPE 3791 presented at the 1972 Improved Oil Recovery Symposium, Tulsa, OK, April 16–19.

9. Gogarty, W. B., H. P. Meabon, and H. W. Milton, Jr. 1970. "Mobility Control Design for Miscible-Type Waterfloods Using Micellar Solutions," *J. Pet. Tech.* pp. 141–147.

10. Matthews, C. S., and D. C. Russell. 1967. *Pressure Buildup and Flow Tests in Wells,* Monograph Series. Society of Petroleum Engineers of AIME, Dallas, *1.*

11. Earlougher, R. C., Jr. 1977. *Advances in Well Test Analysis,* Monograph Series. Society of Petroleum Engineers of AIME, *5.*

12. Gogarty, W. B. 1983. "Enhanced Oil Recovery Through the Use of Chemicals, Part 2," *J. Pet. Tech.*, pp. 1767–1775.

13. Trantham, J. C., et al. 1977. "North Burbank Unit Tertiary Recovery Pilot Test," Second Annual Report, May 1976–May 1977, DOE BERC/TPR-77/5.

14. Trantham, J. C., et al. 1980. "North Burbank Unit Tertiary Recovery Pilot Test," Final Report, DOE/ET/, 13067-60.

15. Goodlett, G. O., M. M. Honarpour, H. B. Carroll, and P. S. Sarathi. 1986. "Lab Data Aid Design and Evaluation of Polymer Floods—Part 2," *Oil & Gas J.*, pp. 82–90.

16. Goodlett, G. O., M. M. Honarpour, H. B. Carroll, and P. S. Sarathi. 1986. "Critical Parameters for EOR Techniques are Reviewed—Part 3," *Oil & Gas J.*, pp. 59–63.

17. Goodlett, G. O., M. M. Honarpour, H. B. Carroll, and P. S. Sarathi. 1986. "Diverse Mechanisms Add to Increased Oil Production in Thermal and Gas Projects—Part 4," *Oil & Gas J.*, pp. 98–102.

18. Milton, H. W., Jr., P. A. Argabright, and W. B. Gogarty. 1983. "EOR Prospect Evaluation Using Field Manufactured Polymer," paper SPE 11720 Presented at 1983 California Regional Meeting, Ventura, CA, March 23–25.

19. Gogarty, W. B. 1967. "Rheological Properties of Pseudoplastic Fluids in Porous Media," *Soc. Pet. Eng. J.*, pp. 149–160.

20. Gogarty, W. B. 1967. "Mobility Control with Polymer Solutions," *Soc. Pet. Eng. J.*, pp. 161–173.

21. Dreher, K. D., and W. B. Gogarty. 1979. "An Overview of Mobility Control in Micellar/Polymer Enhanced Oil Recovery Processes," *J. of Rheology*, pp. 209–229.

22. Christiansen, R. L., and H. Kim. 1984. "Rapid Measurement of Minimum Miscibility Pressure Using the Rising Bubble Apparatus," paper SPE 13114 presented at the 1984 Annual Meeting, Houston, TX, September 16–19.

23. DeHekker, T. G., J. L. Bouzer, R. V. Coleman, and W. B. Bartos. 1986. "A Progress Report on Polymer Augmented Waterflooding in Wyoming's North Oregon Basin and Byron Fields," paper SPE 15179 presented at the 1986 Rocky Mountain Rendezvous, Billings, MT, May 18–21.

24. Nigra, S. S., J. P. Batycky, R. E. Neiman, and J. B. Bodeux. 1986. "Stability of Waterflood Diverting Agents at Elevated Temperatures in Reservoir Brines," paper SPE 15548 presented at the 1986 Annual Technical Conference and Exhibition of the SPE, New Orleans, LA, October 5–8.

25. Gogarty, W. B. 1976. "Status of Surfactant or Miscellar Methods," *J. Pet. Tech.*, pp. 93–102.

26. Exxon Corp. 1982. "Improved Oil Recovery."

27. Gogarty, W. B., and W. C. Tosch. 1968. "Miscible Type Waterflooding: Oil Recovery with Micellar Solutions," *J. Pet. Tech.*, pp. 1407–1414.

28. Davis, J. A., Jr., and S. C. Jones. 1968. "Displacement Mechanisms of Micellar Solutions," *J. Pet. Tech.*, pp. 1415–1428.

29. Healy, R. N., and R. L. Reed. 1977. "Immiscible Microemulsion Flooding," *Soc. Pet. Eng. J.*, pp. 129–139.

30. Gogarty, W. B., and J. A. Davis, Jr. 1972. "Field Experience with the Maraflood Process," paper SPE 3806 presented at the 1972 Improved Oil Recovery Symposium, Tulsa, OK, April 16–19.

31. Bragg, J. R., et al. 1982. "Loudon Surfactant Flood Pilot Test," paper SPE 10862 presented at the 1982 SPE/DOE Enhanced Oil Recovery Symposium, Tulsa, OK, April 4–7.

32. Doll, T. E. 1986. "An Update of the Polymer-Augmented Alkaline Flood at the Isenhour Unit, Sublette County, Wyoming," paper SPE 14954 presented at the 1986 SPE/DOE Symposium on Enhanced Oil Recovery, Tulsa, OK, Aprirl 20–23.

33. Shell Development Co. 1986. "Enhanced Recovery Pilot Begins in Louisiana," Houston.

34. Graue, P. J., and T. R. Blevins. 1978. "SACROC Tertiary CO_2 Pilot Project," paper SPE 7090 presented at the 1978 SPE Symposium on Improved Methods of Oil Recovery, Tulsa, OK, April 16–19.

35. Kane, A. V. 1979. "Performance Review of a Large Scale CO_2 Wag Project SACROC Unit-Kelly Snyder Field," *J. Pet. Tech.*

36. Stalkup, F. I. 1983. *Miscible Displacement*, Monograph Series, Society of Petroleum Engineers of AIME, New York, 8.

37. Suffridge, F. E., D. W. Dauben, and J. A. Pezzullo. 1986. "Evaluation of the Little Knife CO_2 Minitest," U.S. DOE, Bartlesville, OK.

38. Saner, W. B., and J. T. Patton. 1983. "CO_2 Recovery of Heavy Oil: The Wilmington Field Test," paper SPE 12082 presented at the 1983 SPE Annual Technical Conference, San Antonio, TX, October 5–8.

39. Prats, M. 1982. *Thermal Recovery*, Monograph Series. New York: Society of Petroleum Engineers of AIME, 7.

40. Chu, C. 1983. "State-of-the-Art Review of Steamflood Field Projects," paper SPE 11733 presented at the 1983 Annual California Regional Meeting, Ventura, CA, March 23–25.

41. Matthews, C. S. 1983. "Steamflooding," *J. Pet. Tech.*, pp. 465–471.

42. Chu, C. 1981. "State-of-the-Art Review of Fire Flood Projects," paper SPE/DOE 9772 presented at the 1981 Joint SPE/DOE Symposium on Enhanced Oil Recovery, Tulsa, OK, April 5–8.

43. Matheny, S. L., Jr. 1980. "EOR Methods Help Ultimate Recovery," *Oil & Gas J.*, pp. 79–124.

44. Leonard, J. 1982. "Steam Dominates Enhanced Oil Recovery," *Oil & Gas J.*, pp. 139–159.

45. Leonard, J. 1984. "EOR Set to Make Significant Contribution," *Oil & Gas J.*, pp. 83–105.

46. Leonard, J. 1986. "Increased Rate of EOR Brightens Outlook," *Oil & Gas J.*, pp. 71–101.

47. Gogarty, W. B. 1978. "Micellar/Polymer Flooding—An Overview," *J. Pet. Tech.*, pp. 1089–1101.

48. Gogarty, W. B. 1983. "Enhanced Oil Recovery Through the Use of Chemicals—Part 1," *J. Pet. Tech.*, pp. 1581–1590.

49. Bragg, J. R., R. E. Rosener, and J. E. Strassner. 1982. "Measuring Well-Injection Profiles of Polymer-Containing Fluids," paper SPE 10690 presented at the 1982 SPE/DOE Enhanced Oil Recovery Symposium, Tulsa, OK, April 4–7.

50. Knight, B. L., and M. J. Davarzani. 1986. "Injection Well Logging Using Viscous EOR Fluids," *SPE Formation Evaluation*.

51. Perry, C. W., et al. 1979. "The Status of Enhanced Oil Recovery in the United States with an Overview of the U.S. Department of Energy Program," *Proc.*, 1979 World Pet. Cong., Bucharest, September 9–14.

52. Mills, A. 1980. "Economic Incentives Get Better and Better," *American Oil & Gas Reporter*, pp. 84–88.

53. Ittig, J. 1982. "New WPT Considerations for Tertiary Recovery," *Oil & Gas J.*, pp. 140–148.

54. Gallagher, T. J., Jr. 1981. "Tertiary Recovery and the New 25% R&D Tax Credit Under the 1981 Economic Recovery Act," *Oil & Gas J.*, pp. 113–120.

Part II

Small-Scale Hydropower Programs in Developing Countries

7

Overview

For a long time the energy contained in water runoff has been beneficially exploited. Initially this was accomplished by direct conversion into mechanical energy, but with the advent of electricity about a century ago, conversion has shifted predominantly to electricity generation to take advantage of this flexible and versatile energy form. Early hydropower schemes were small in size to conform to the relatively small power markets available. It was soon realized, however, that considerable economies of scale could be achieved by exploiting larger river flows. This realization went hand in hand with the development of electric power transmission, which permitted large and dispersed power markets to be supplied from a single scheme. Smaller schemes tended to fall out of favor because their relatively minor contribution to the demand coverage did not seem to justify the financial and infrastructural efforts involved to bring them to fruition.

This situation changed radically when fossil fuel prices began to escalate in the early 1970s and greatly increased the economic attractiveness of small-scale hydropower, especially in the developing countries where the institutional and financial situations and the lack of adequately large power markets tend to inhibit the development of many of available main river schemes. It is evident that adequate supplies of energy at an acceptable price and, as far as possible, from domestic sources would greatly encourage economic progress in rural and outlying areas. This has led a number of developing countries to look at their available water resources and consider how they might best be exploited, with the objective of improving rural electrification for the benefit of the resident population. One approach for the establishment of programs for the development of small-scale hydropower in the developing countries is discussed in this section. Its aim is to put forward guidelines to help potential developers in setting up such programs.

Before considering the detailed issues involved, it is necesary to define the meaning of "small scale." Size definition is usually arbitrary and is often related to the size of the interconnected power system to which a new scheme is to contribute. For an established power network, a contribution of 1% or less is generally considered small. The majority of small-scale hydroplants will, however, feed isolated load centers or very small existing networks, and a relative size definition is then more difficult to apply. As a broad generalization, the size of a scheme considered "small scale" here would lie between about 500 kilowatts (kW) and 10 MW, but both the lower and upper limits of this range are arbitrarily chosen. For the purpose of this section, it is assumed that plants within this range built in a developing country would require technical and managerial support from sources external to the area in which the plant is to be built. This support would be obtained either from national or expatriate sources depending on the facilities and experience available. Below 500 kW, the scheme can be classified as a "micro" development, which, for reasons of cost alone, will have to rely greatly on local support for planning and design. Above about 10 MW, the investment requirements alone make it necessary to mount an extensive development effort with considerable reliance on external assistance. The many aid and technical assistance programs available to developing countries do, however, tend to cut across such considerations. Development assistance is often made available in support of the social objectives that a new power supply is intended to achieve and is not directly related to the size of the facilities to be provided.

THE ROLE OF SMALL-SCALE HYDROPOWER IN THE NATIONAL ENERGY PLAN

The mechanical energy produced by hydropower can be used directly for small industrial processes such as flour milling and wood cutting or for water pumping, but these are of very limited scope. The only hydropower application of universal importance is for the production of electricity whereby the water runoff provides the source of primary energy. It is with this application that the term "hydropower" is conventionally linked and thus becomes an element in an electricity supply scenario. Small-scale hydropower becomes an option in cases where there is a demand for an electricity supply on a small scale, either in close proximity to the hydropower source or within an economically and technically acceptable radius from it. The hydropower source is formed by a naturally replenished river course having an appropriate gradient or drop in height.

In the context of national energy planning, small-scale hydropower generation is in practice restricted to rural and outlying areas not conveniently or economically served from extended electrical networks. There are exceptions in cases where small hydro schemes were built many years ago and were overtaken in scale by larger projects but are still in service to contribute to an interconnected network supply. There are also cases where small schemes can usefully be linked with such a network, for example, to long and weak transmission lines extending from larger but distant generating plants where the small scheme can help to maintain electrical stability. In general, however, the large demands for electricity at the principal load centers and the economies of scale of the larger plants have displaced small-scale generation away from central locations to the outlying areas. Development of small-scale generation thus becomes dependent on the policy adopted for the electrification of these areas.

Although the energy supply for the central areas—the economically important areas—in the developing countries has tended to absorb the major planning effort, small-scale schemes were planned and built in those outlying areas that could offer an adequate incentive to meet local social and economic needs and for which investment funds could be found. These outlying centers generally formed a nucleus for small-scale development of industry, agriculture, or commerce and often attracted or contained a relatively large population. Local power schemes were mostly based on diesel plants, even in areas where exploitable hydro resources were available. Diesel units are relatively cheap, easy and quick to procure and install, simple to operate, and flexible in operation to cope with the often greatly variable demand characteristics of newly developed local centers. Diesel fuel has now become expensive, especially in remote areas where delivery costs can be high. Imports of the fuel into countries deficient in domestic oil resources cause a drain on foreign currency reserves. Problems have also arisen over the maintenance of diesel engines and the procurement of spares for them. Energy planners are therefore looking for alternative—and preferably local—sources that could gradually replace diesel power production and take over the expansion of electricity supply in the rural areas where, because of the small and scattered markets, supply costs are inherently high.

Hydropower is such a resource and can offer an economic way of taking over—wholly or partially—the service performed by diesel plants. The benefits from a local hydro scheme do not necessarily accrue to the local electricity consumer alone; the scheme can bring wider benefits for the national economy as a whole by:

Reducing the drain on foreign currency reserves where hydropower can replace diesel power.

Reducing deforestation where the electricity produced can replace firewood.

Reducing the division between urban and rural life and the social stresses this produces.

Discouraging population drift to urban areas through the improvement of living conditions of the rural population.

Therefore there is now considerable interest in promoting small-scale hydropower as an element in the national energy plan, primarily in support of rural electrification on economically attractive terms. The small and variable power demands and the consequent size and characteristics of the facilities needed to meet them tend to make small schemes expensive, whatever the source of the power. This has to be kept in mind when setting up a rural electrification scenario and it must be weighed against the social and economic benefits that rural electrification can bring. Purely commercial benefit and cost considerations should not be the determinant in assessing the role of small-scale hydropower in supporting social and economic progress in a developing country and in deciding on the priority to be accorded to it in the energy plan.

THE POTENTIAL FOR SMALL-SCALE HYDROPOWER

A small scheme has to be competitive with alternative power sources in the price of the electricity delivered from it to the consumers. This means that hydro projects have to be sought out that can be developed cheaply and quickly, using a readily exploitable head—generally not below about 10 meters—and a reasonable water flow produced by either a waterfall, a rapidly flowing river, or small lakes and reservoirs located close together. Artificially created storage reservoirs can greatly improve the value of the electricity produced, but they are usually too expensive to justify inclusion in a small scheme. Seasonal variations in river flow are unavoidable in most parts of the world. However, the river selected for development should nevertheless carry an exploitable amount of water for as long an annual period as possible and should not give rise to excessive flood conditions.

Where the hydro scheme supplies electricity to an isolated load, it should be able to meet as much as possible of the power requirements of that load throughout the year. Any shortfall in the energy it can deliver to the load will have to be made up from another plant, generally diesel, either specifically installed to supplement the hydro output or relocated from an existing supply in cases where hydro production is intended to replace diesel generation. Where the hydro scheme is connected to a network also supplied from other power sources, including the thermal power plant, its output should be maximized at all times to save thermal generation. If the scheme contains storage, its production will have to be coordinated with other plants in the network so that maximum benefit can be derived from the combined operation.

The location of the scheme with respect to the isolated load center or the nearest infeed point to the network is crucial for its success because the costs of any required transmission link will form part of the expenditure on the scheme; excessive transmission costs can invalidate economic merits. The burden imposed by the transmission costs will be particularly heavy if, because of lack of water runoff, the output from the scheme falls seriously short of the load requirements.

Hydropower development necessitates extensive investigations (described in this section) before a firm decision on implementing a given scheme can be made. Every scheme will have its own specific site characteristics and will therefore require its own series of investigations. The organizational and financial issues involved may discourage serious consideration to small and scattered possibilities capable of making only a minor contribution even to small rural energy markets. Developing countries can obtain help from a number of agencies for establishing small-scale hydropower programs and carrying out the necessary preinvestment studies. The planning work will be greatly facilitated if several projects can be developed together or in sequence, although the amount of information that can be transferred from one scheme to another will usually be small. In spite of its strictly local impact, small-scale hydropower can confer appreciable benefits to the rural economy by way of low-cost electricity supplies from indigenous sources, and this can justify a considerable effort for building a scheme.

ADVANTAGES AND LIMITATIONS OF HYDROPOWER

The water runoff on which hydropower is based is a renewable resource, generally dependent only on the precipitation in the catchment area. Since the precipitation pattern is subject to considerable changes from day to day, season to season, and year to year, the water runoff can vary within wide limits in a largely unpredictable way. In consequence, the dependable or firm output with which a hydro scheme can be credited is often only a small fraction of the total output that the scheme can produce over a given period of time. The dependable output will, in any case not be the same amount from year to year. The uncertainties are greatest with a run-of-river scheme in which there is no water storage upstream of the power plant. Firming-up of the output, with the objective of increasing the dependable capacity, requires storage. This can be costly and environmentally unattractive but can greatly enhance the value of the scheme because the output will become more predictable and can be more readily fitted to the demand pattern that the hydro plant is to supply. But whatever the type of scheme selected, the hydrological conditions governing its performance will remain unpredictable to some extent.

Other events can also affect the water runoff over the long term, in particular:

Sediment of sand, soil, mud, and stone carried down by the river and its tributaries can affect the river flow—especially during dry periods—and also gradually reduce the capacity of any storage reservoir built upstream of the power plant; sediments can also cause erosion of the turbine blades.

Major changes in the river flow regime can be brought about by catastrophic changes upstream, especially during flood conditions.

These matters make it essential to carry out a thorough analysis of the hydrology and topography in the catchment area before a commitment to build the hydro scheme is finally made. The extent of the risk involved depends largely on the size of the scheme envisaged and consequently on the amount of the required investment.

Nevertheless, hydro schemes have important advantages over other means of power generation, apart from the renewable nature of the energy they produce:

They do not cause pollution.

They do not cause any loss of water; water discharged from the power plant is normally fully restored to the river system from which it was originally taken.

They involve well-known and fully proven technology.

Their electro-mechanical equipment has by now become largely standardized, for small-scale plants at any rate, and it is therefore cost-efficient.

Their civil engineering component involves relatively simple skills and can offer employment to local labor.

The schemes are simple and cheap to operate and maintain; they require only a small labor component for operation and lend themselves to a high degree of automation.

The schemes have a long service life.

They can be developed in conjunction with water uses for irrigation and water supply and can thus benefit from sharing investment costs with these purposes.

Hydro schemes do, however, have limitations beyond those imposed by the hydrological conditions.

They require extensive and consequently costly investigations before a commitment for their construction can be made.

They are capital-intensive even though much of their civil works can be built with local resources.

They modify the river flow between the point of water abstraction and the point of discharge and this can have an adverse effect on the local environment and ecology; the effect becomes more serious if any storage is incorporated in the scheme and the river flow downstream is thus affected for long periods of time.

They are strictly site-dependent and might therefore be located relatively far from the load centers they are to supply; interconnection with the load centers may require extensive transmission lines, which can greatly increase the cost and reduce the effectiveness of a given scheme.

They may be out of operation for parts of the year—during the dry season—when the water flow through the turbines is too low to permit power generation.

They are relatively inflexible once they are built; unless their design allows it, they cannot be adapted to changes in the demand pattern nor can they be extended.

Their long asset life brings with it the danger of obsolescence of their equipment and difficulties in obtaining spare parts.

These limitations do not necessarily turn the scales against hydropower, but they have to be clearly recognized and dealt with in the planning stages of a project. Planning demands appraisal of the operational problems that the hydro scheme is likely to encounter throughout its lifetime and the establishment of design criteria that can overcome these problems in the most effective way.

8

Planning

POWER MARKET DETERMINATION

Building a power plant of whatever size can be justified only if there is a market for its output. This market can be formed by:

Natural load growth in an existing system.
Replacement of an obsolete facility or replacement of an existing power plant, say a diesel station, by a hydro plant.
A new demand resulting from the electrification of a new load center or new industrial establishment.

In each case, the capacity and operating regime of the new station must be designed to meet the demand assigned to it in the best possible way, alone or together with other power stations feeding the same network.

The electricity demand of any group of consumers can change from moment to moment in a largely unpredictable way. Demands vary between peaks when high consumer loads coincide and troughs where low consumer loads are dispersed in time. The relation between the coincident consumer demand and the moment in time when this demand occurs is expressed by the load-duration curve of the network to which the consumers are connected. This demand characteristic is usually specific for this network and is not necessarily valid for any other network, even in the same climatic zone where both consumers and their consumption patterns have common features. The load duration determines what output will be required from the hydro plant at what time, but this pattern will not be static. It will vary in a random way within certain limits that usually become narrower as the demand becomes larger. The maximum demand for which a hydro scheme is to be designed and the load variation it has to accept have to be determined in the planning stages of a project, but the situation is complicated by the fact that every hydro scheme has its own output limitations that are governed by hydrology and unrelated to the power market requirements. It is therefore often difficult for a run-of-river scheme to match the load variations of a given consumer group, and compensation from either hydro storage or supplementary thermal power may then have to be provided to bring the hydrological producibility into balance with the variations of electrical demand. The situation will be most difficult for an isolated scheme where complementary energy supplies are not available. A load allocation analysis will determine to what extent the new hydro plant can contribute to the system demand. This analysis is usually undertaken for average hydrological years and sensitivity studies that explore the likely situation under flood and drought conditions. Such an analysis is important because arbitrary selection of design parameters for a new hydro scheme can easily lead to inefficient utilization of the plant.

The electricity demand forecast has to be projected well beyond the time when the new scheme is to become operational. How far it should extend into the future will depend largely on the size and precise role the scheme is to play in a given power supply scenario. The demand pattern will always be dynamic, that is, it will change with time. Where a hydro scheme is intended to supply an existing network, however small, the demand it will have to meet will not necessarily be only that available at the time of its interconnection with the network; rather it will be a growing demand to which the scheme will have to contribute according to its share of the load. This is an important consideration for the dimensioning of the scheme.

Electricity demand growth is caused by three events:

1. Growth in consumption of the existing consumers.
2. Supply of new consumers in the network.
3. Extension of the network to new supply areas.

Network extensions do not necessarily raise the total demand of an area, but they do raise the demand in the particular network that is being extended. Demand analysis must always be restricted to interconnected loads and must not include loads that will remain in isolation.

The demand in an existing network will have been monitored over some years and its characteristics and rate of growth will be known. In most networks, the three events noted above will occur together and it will be difficult to isolate each component and analyze it separately from past load statistics. It is usual then to extrapolate the historical load growth experience into the future but to correlate this with other factors that may have a bearing on the electricity demand and on the load characteristics. The principal factors to be taken into account are:

1. Electrification program.
2. General development plans for the supply area.
3. Economic indicators, both national and local, such as the domestic products, industrial activity, housing developments, population growth, and movement in and out of the supply area, and employment statistics.
4. Availability and usage of alternative energy forms (solid fuel, oil, gas, firewood, and agricultural waste) and their potential replacement by electricity.
5. Electricity consumption statistics broken down into consumer categories and numbers, utilization patterns, and specific consumption per capita.

6. Load curves for the principal consumer categories (domestic, commercial, industrial, agricultural, institutional, and governmental).

Judgment factors can then be derived from analyses of the parameters that have a bearing on the demand. These factors are then used to adjust and weight the demand characteristics extrapolated from historical experience to arrive at probable figures for the demand reached at a given time. "High" and "low" load growth scenarios are established in this way, with the most probable load figures lying in the zone between these two limits. This zone will become wider as the timeframe for the demand prediction is extended. Since the uncertainty of the demand forecast will increase with the length of its projection, estimates going beyond 10 to 12 years are of little practical value for project planning purposes.

The "high" growth scenario indicates the earliest time, and the "low" growth scenario the latest time, by which a new power plant will be needed. The gap between these two limits shows the maximum time period by which the plant would be antedated in case of an unfavorable load growth experience. Judgment can then be brought to bear to decide on the appropriate installation date for the plant.

A different approach is necessary when a new and isolated load center is to be supplied. Demand analysis must then be based on synthesis and aggregation of potential loads. The task is relatively simple for the supply of an industry where the potentially connected load, and its anticipated utilization pattern, can offer a reliable guide. A more searching investigation will be necessary for the supply of a new load center comprising a predominantly domestic and commercial element. The climatic conditions and social circumstances of the potential consumers need to be investigated and their likely electrical consumption and usage patterns identified. Help can sometimes be obtained from the experience of already electrified areas located in a milar climatic, economic, and social environment and containing a similar spectrum of consumers. Canvassing of potential consumers can also help, but care in interpreting the results is necessary because potential consumers are unlikely to have a realistic appreciation of their electricity requirements.

Typical load figures for households in tropical and subtropical regions of developing countries amount to about 200 to 500 watts and up to 2 kW for large consumers. The longer term growth pattern will also be much influenced by the electricity tariff and by the availability and price of alternative energy forms—firewood and kerosene, for example—which consumers may prefer because they are more familiar with them.

The power market for any particular scheme must lie within a radius that can be economically reached by the network supplied from this scheme. The market will therefore be physically restricted by the extent of this network. This needs to be kept in mind when considering hydropower development in rural and outlying areas.

THE PROJECT CYCLE

The project cycle extends from the identification of a potential project to its final commissioning for the supply of power. It can be divided into two components:

1. The preinvestment phase, which includes the planning, development, and design work required to prepare the project for construction.
2. The investment phase, which includes the contracting procedure and the construction process as such up to the point of commissioning; in some cases it can also extend over a period initial operation.

Each of these phases can be divided into several successive activities that are to bring the project to its final completion.

Preinvestment Phase

The extent of the preliminary work to be done in this phase depends greatly on the quality of the available information and on the ease with which it can be put together and processed to support a serious development proposal. The work is usually divided into five successive activities that are to establish an increasingly reliable technical and economic justification for the project, provide guidelines for project execution, and permit firm investment decisions to be taken. The sequence of activities is as follows:

1. Project identification: Finding projects, from the study of maps and surveys on the ground, which would contribute to the requirements of an existing or potential power market. The aim of this exercise is to identify projects that might be potentially worth exploiting.
2. Initial survey: Determining the principal characteristics of such projects and deciding which of those already identified may be worth investigating further. The survey campaign often leads to the setting-up of an inventory of such projects for a given area or region, with priority assigned to projects in accordance to their potential merit.
3. Data collection: Collecting information on the physical features of the projects identified as being of adequate merit, on their potential design parameters, and on the physical and economic environment within which they will have to function. Assessment of the physical features will include the hydrological and topographical conditions under which the project will have to operate, the characteristics of the construction site both above and below ground and of the access to them, the power markets available to the projects, and the features of the network to which they could be connected.
4. Prefeasibility study: Reviewing the technical and economic aspects of a specific proposed development to demonstrate that it is of sufficient merit to warrant further work. The study explores the physical characteristics, the size and potential output, and the available power markets. Outline designs are prepared and costed, and first indication of economic merit is established. It should be possible to determine at this stage, whether a given project proposal justifies proceeding with its development and to identify additional data that may be needed to firm up the information on it.

5. Feasibility study: Preparing a detailed and final analysis of the project usually in a form suitable for an investment decision to be taken (i.e., for the preparation of a "bankable" report). This study repeats in essence the work done in the prefeasibility phase but with a more detailed and specific approach and to a significantly greater degree of reliability. Project design and costing should be in sufficient detail to provide a sound basis for ultimate construction, although further refinements may have to be introduced before contract can be let. The economic appraisal should be adequate at this stage to permit a final investment decision to be taken. The influence of uncertainty in the principal project parameters on the merit of the proposal will have to be tested by sensitivity analyses. The quality and extent of the information produced should be able to satisfy potential financing institutions.

6. Project design (detail engineering): Preparing final designs and establishing specifications for initiating the tendering procedure with potential contractors and manufacturers and also setting out a detailed work program and for the ultimate construction of the scheme. This is still a preinvestment activity, but it should not be undertaken unless project implementation has been decided on and the investment problems have been cleared.

7. Financial report: Preparing a detailed economic and financial analysis of the project beyond the information given in the feasibility study. This is sometimes required by the financing institutions to confirm that the project is viable. The analysis includes specific financing proposals, schedules of cash flows during construction, income and expenditure forecasts for operation, and the proposed organization of the undertaking owning and managing the scheme.

This sequence of activities will have to be adapted to the particular circumstances under which the project is to be built. For small-scale schemes, the preinvestment activity will have to be greatly simplified to keep within the framework of available funds. The economic and social merit of the project needs to be established as early as possible in the project cycle, preferably during the initial survey. If the merit is sufficiently great, the funds that can be devoted to the preinvestment activity and consequently the depth and extent of the preinvestment investigations can be greater, but they must still stand in a reasonable relation to the overall size and cost of the proposed scheme. This point is of particular significance for data acquisition, which in many developing country situations cannot be comprehensive without an unduly protracted and correspondingly expensive campaign.

Two potential pitfalls must also be kept in mind in the preinvestment phase:

1. The danger of the effort becoming too elaborate and extensive in relation to the size and value of the scheme, particularly if the volume of the investigations undertaken is not closely controlled.

2. The danger of the effort leading to inconclusive and repetitive studies if firm decisions are not taken at the appropriate point in time and if development proposals become unduly deferred.

This can result not only in excessive costs of the preinvestment effort but also to an accumulation of rapidly outdated studies. The costs of the preinvestment effort devoted to small-scale schemes should not exceed about 10–12% of the total investment required for implementation. If this range is substantially exceeded, the total investment needs will become inflated and the economic feasibility of the project will be seriously put into question. Where some part of the preinvestment expenditure and the balance needed for completing the work should be appropriately reduced.

It is difficult to set out general guidelines for rationalizing the preinvestment effort because every case will impose different conditions. The task of data collection can often be shared with other activities and its time requirement reduced. The feasibility and design studies will make up the most costly part of the exercise. Considerable economies can be achieved by combining the prefeasibility and feasibility studies into one activity that:

- provides a firm foundation for the technical feasibility of the project
- presents cost estimates and economic appraisal with sufficient detail and reliability to satisfy the financing institutions
- develops the outline design far enough to permit specifications for the civil works and power plant to be based on it

All parties involved in the project will have an interest in ensuring that costs are kept in check without sacrificing the soundness and viability of the proposed scheme.

Investment Phase

The construction procedure can be initiated once specifications for the principal items of the project are available and financing is secured. The essential elements of the construction phase are the following:

1. Tendering procedure: Inviting selected contractors (for the civil works) and manufacturers (for the electro-mechanical equipment) to tender to the specifications supplied to them, evaluating the tenders, and placing appropriate contracts. It is advisable in the interest of economy to restrict the number of contracts placed to a minimum, preferably only two—civil works and electro-mechanical equipment. Local contractors directly employed by the owner of the plant may well be available, but expert supervision would probably be needed in such cases.

2. Construction: Construction proceeds in accordance with the work program previously laid down and modified in consultation with the contractors and manufacturers. Arrangements for an adequate level of technical supervision will be essential. Civil works will usually be built first and followed by the installation of the electro-mechanical equipment. Familiarization of the ultimate operator with all aspects of the construction will be essential.

3. Commissioning: On completion of construction, start up and acceptance trials will be carried out to test the functioning of the whole system and ensure that it complies with the requirements originally laid down in the specifications. This is followed by formal commissioning when the plant is handed over for normal commercial service.

The financing institution involved will want to be kept informed of the detailed cash flow throughout the investment phase and an appropriate reporting procedure will have to be set up. It will also be necessary for the owner of the plant to prepare for operation and maintenance after commissioning by setting up an operation and maintenance organization that can recruit and train suitable personnel, in close liaison with the contractors and manufacturers employed on the project. The owner should also establish a sales organization that will be responsible for revenue collection and contact with the potential consumers. Realization of the economic targets set for the scheme in the planning stages depends critically on the integration of the output from the scheme into the previously established power market and on securing an adequate revenue from this market.

POWER SYSTEM INTEGRATION

The output from a small hydro scheme will be used to supply electricity to an existing network or to an isolated group of consumers, occasionally even to a single consumer of adequate size. Interconnection with a network that also draws power from a thermal power plant has the advantage that stream flows arising during hydrological peak flow periods can often be utilized to save thermal generation; they thus become firmed up by the thermal power plant and the value of the hydro energy is increased through this firming-up process. During periods of low water runoff, the thermal plant can supplement output deficiencies from the hydro plant. Such complementary operation has considerable advantages because it permits better use of the hydro output and allows load-following to be achieved by the combination of thermal and hydro power productions without either having to restrict the hydro output during periods of high runoff or accept a shortfall of available energy during the dry season.

Integration of a hydro plant into an existing power system requires careful analysis of the role each contributing plant is to play in meeting the system demand in an optimum way. The analysis is essential when the installed capacity, and in appropriate cases the storage volume, of a new hydro scheme have to be decided. The operational flexibility provided by an interconnected power system allows hydro plants to take advantage of economies of scale, of simpler plant arrangements, of reduced stand-by provision, and of cheaper designs—for example, it allows the use of induction generators. Integrated operation can also facilitate maintenance; hence this can be made a system responsibility instead of an isolated task for each power station. The advantages of system integration can encourage interconnection of hydro schemes originally conceived for isolated operation.

When the introduction of a hydro plant into a mixed hydro-thermal system is considered, system studies will have to explore the:

Influence of the new scheme on the operation of the existing system and on the structure of its production costs.

Effect of the new scheme on the expansion of generation and transmission facilities.

Optimum dimension of the new scheme in relation to system requirements and the phasing of its development.

Transmission capacity and any strengthening of the network needed for absorbing the output from the new scheme.

Such studies are essential to ensure that the new scheme can satisfactorily fit the purpose for which it was originally conceived and that it is fully compatible with system requirements.

System integration can also be achieved, hydrologically speaking, by incorporating the hydro power component into a water supply or irrigation project. Advantages are then derived mainly from a sharing of investment costs between the different purposes. Water supply and irrigation schemes generally allow low-head production only, but much progress has been made in designing turbines for low-head conditions, and a number of interesting concepts are now available. Such machines are generally of small size and very suitable for the scale of development likely to be required in remote locations and for use in association with the multipurpose exploitation of water resources. The operating regime of the hydro component in a multipurpose arrangement is generally dictated by the use of the water for the other purposes, and hydro power is thus produced as a by-product rather than in its own right. Studies must then be carried out to ascertain how this type of hydro power production can be best fitted to a given demand pattern. The problem is greatly facilitated if the hydro component supplies energy to an interconnected network that can make up differences between the hydro output and the load requirements.

ENVIRONMENTAL ISSUES

A hydro scheme can make a significant impact on the environment. This is caused primarily by the extensive civil engineering works involved and by the changes to the flow of the river that the scheme brings about. The extent of the environmental encroachment depends largely on the size, design, and layout of the scheme. The civil engineering works can be divided into three principal components:

- water impoundment
- headrace and pressure conduits
- powerhouse with tailrace

Each of these components will make its own particular impact. The water impoundment arrangements are often located in relatively remote and inaccessible places, and although some visual impact cannot be avoided, the disturbance they create can often be reduced by landscaping and suitable treatment of the structures. Pressure conduits and powerhouses can be undergrounded to remove their impact altogether, but this is an expensive process and is not usually feasible for a small scheme. Some reduction of visual impact can also be achieved by surface treatment, but a tailrace located on the surface will be difficult to disguise. The transmission lines emanating from the power plant will face the same environmental problems as transmission lines in general; they often form the most visible part of the small hydro scheme.

Perhaps the most contentious part of the scheme will be the effect of water abstraction on the river flow. The abstracted water passing through the power plant will in most cases be fully restored to the river downstream of the plant and there will be little overall loss of water, but the section between the water intake and the point of discharge back into the river usually has a greatly reduced flow of water and this can seriously impair fish life in the river and local amenity. This effect will be more serious if the abstracted water is transferred to another catchment and not restored to the river in question. If the hydro scheme has significant storage incorporated in it, the river flow regime may be greatly changed and may again cause undesirable loss of water at critical times, especially during the dry periods.

In many cases, environmental disturbances caused by the civil engineering works may be greatest during the construction phase. Although such disturbance is generally much reduced once construction is completed, some residual effects may be left and may not allow full return to the previous undisturbed condition.

Measures to reduce the environmental impact can be costly and can in some cases greatly effect the economic merit of the proposed development. The industrialized world is suffering increasingly from this experience and there can be no doubt that developing countries will also in due course have to consider whether the environmental impact—even in remote areas—is acceptable. It is usual now to undertake an environmental analysis in which the likely effects of the scheme on water, ground, climate, biosphere, landscape, and local land use are investigated. Conditions usually considered are:

1. Undisturbed environment, i.e., the ambient condition.
2. Impact during the construction phase.
3. Environmental effects during normal operation.
4. Environmental impact of breakdowns of the scheme.

The detailed subjects studies are impacts on:

> hydrology
> ecology
> local planning
> visual disturbance of the landscape

air, climate, noise, and vibration
living conditions and quality of life
local activities and employment

Not all these impacts are necessarily negative. In many cases, the pressing need for energy from indigenous sources, lower population densities, and a more relaxed attitude to environmental issues have created a favorable climate for the acceptance of hydropower. Nevertheless, an indiscriminate attack on the local economy and ecology should be discouraged. The extent of the potential disturbance needs to be appraised in the planning stages and measures to contain it included in the development plan. Assistance may have to be given to local institutions to foster awareness of environmental issues and to encourage them to develop measures for reducing environmental impacts. Training may have to be given in methodologies of impact assessment and control. Such measures may go a long way to alleviate opposition to power development, which has become such a significant feature in many parts of the industrialized world.

SOCIOECONOMIC IMPACTS

The merit of a hydro scheme is conventionally appraised on its directly quantifiable benefits or on the monetary return it can earn on the capital invested in it. Schemes that are likely to fall short of predetermined financial targets are in danger of being discarded. The socioeconomic benefits of such schemes are not usually taken into account, primarily because they are difficult to evaluate in numerical terms and to include in the economic calculation.

The social benefits of an adequate supply arise out of the provision of light, heat, and motive power and the greatly enhanced quality of life this can bring about. The economic benefits arise in two ways:

Directly, through the employment opportunities that construction of a new hydro station can create, both with the scheme itself and with the provision of materials and components for it.

Indirectly, by stimulating the local economy and creating employment opportunities in commercial and industrial activities made possible by the energy supply.

The direct benefits are transient and do not usually extend beyond the construction period of the scheme. The operation of hydro stations is not manpower-intensive and the direct employment opportunities they create are very small. The construction phase does, however, offer useful experience and training in civil engineering skills for local personnel and this can often be put to good use once the scheme is completed.

The inherent benefits resulting from more extensive, or intensive, electrification can be every significant. Newly electrified areas in developing countries experience considerable improvements in the local productive capacity. These can lead to a substantial growth in the monetary economy. This is not likely to figure in the appraisal of a new hydro project because of the difficulties of quantification. In general, the lower the ambient intensity of electrification, the greater is the nonquantifiable benefit that electrification can bring.

The direct benefit that a hydro scheme is likely to achieve in comparison with an alternative development may be reinforced eventually with a quantitative evaluation of the indirect benefits resulting from its socioeconomic impact so that a stronger case can be made in its favor. Scenarios will then have to be set up that allow this impact to be numerically computed for the particular social and economic environment that the scheme is to supply. The determination of socioeconomic impacts, and their quantification, are therefore likely to play a much more important role in the future.

9

Economics and Financing

COST ESTIMATES

Detailed cost estimates are required for determining the economics of the project and arranging financing for it. Where the feasibility investigation is undertaken in two stages, cost estimates are first made to a reasonable approximation for the prefeasibility study and are then refined, on the basis of a more extensive investigation, for the feasibility review. Financing is usually arranged on the costs established at that stage, but estimates are occasionally revised where the final design brings out elements that may have a major impact on the costs of particular components.

Cost estimates for hydro schemes are specific for each project and each site. General data are not necessarily relevant, although they give an indication of the ranges within which component costs may be expected to lie and of the cost trends that might be experienced. Cost trends derived from general data show, for example, that:

Specific costs of hydro plant (per kW installed) decrease with increasing capacity and increasing head.

Isolated schemes are generally more expensive than more accessible projects located near existing power markets.

Use of local materials and equipment and of simple technologies reduces specific costs, but the advantage becomes less marked with larger capacities.

Investment requirements increase with the number of units installed in the station, but this advantage is counteracted by the better availability of the plant.

For small schemes, below about 1 MW, the electro-mechanical component is responsible for the largest proportion of the total costs, but the civil works component becomes more prominent for larger schemes and absorbs an increasing proportion of the total costs (see figure 9.1 for two typical cases).

The most expensive item of the civil works is the headrace and pressure conduit system, and this therefore requires greatest care in design optimization (see figure 9.2).

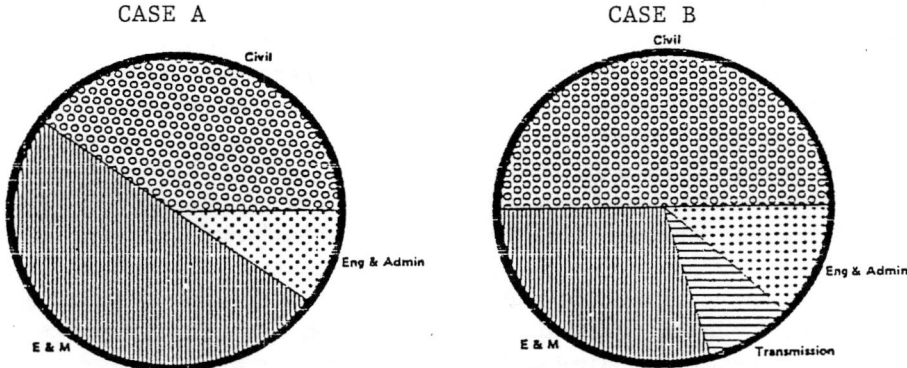

Figure 9.1 Relative scheme costs.

General data are more appropriate for thermal—especially diesel—plants where the influence of site conditions on the project cost is small. If, for example, the economic merit of a hydro and a diesel scheme are to be compared, the diesel solution can be costed reliably from general data and a design study for it need not be undertaken. The hydro scheme, on the other hand, must be costed with reference to the particular site conditions established and the corresponding engineering solution developed. Even so, the estimates are no more than predictions and may not be fully substantiated when final design data or construction proposals are in hand. If there are major discrepancies, it may be necessary to review the economic merit of the scheme at that stage to confirm that it still meets acceptable targets.

Cost estimates must include all expenses associated with the building of a particular scheme, including costs of engineering and project management from the initial surveys to final commissioning of the station, all construction expenses associated with the project, and, where appropriate, all financing charges. All costs must be related to the final installation on site and therefore also include expenses for erection, site personnel and facilities, transportation of materials and equipment to the site, insurance, taxes, duties, and other levies. The costs must make adequate al-

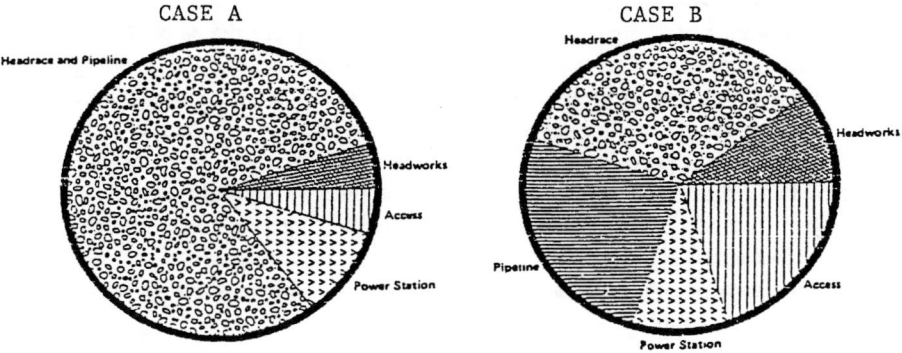

Figure 9.2 Relative civil costs.

lowance for the expenditure in foreign currency on imported equipment, and it is therefore usual practice to state costs in local and foreign currency separately for presentation to the corresponding financing agencies. The terminating points of the project must be clearly defined for costing purposes so that the price of land acquisition, preparation of access, environmental improvements and of the transmission link can be properly taken into account.

The estimates should make adequate provision for contingencies, say 20% for the civil engineering component and 15% for the electro-mechanical equipment; these margins may be reduced where the cost data are judged to more reliable. The data of the estimates and the corresponding price levels must be carefully defined and the contingency margin increased where substantial price escalation to the time of placing the contracts is anticipated. Costs of existing facilities that are incorporated in the new project need not be included in the estimates unless the owner is to receive a payment in compensation. Expenses met by other parties, through aid, technical assistance, or support from government and national institutions must be included for purposes of economic project evaluation but can be excluded from computation of the funding requirements. Shadow pricing enters into the costing process only for the economic evaluation; it is accounted for otherwise by making proper allowance for the commercial and institutional margins on imported materials and equipment. On balance, cost estimates should be as comprehensive and reliable as possible so that difficulties arising from inadequate financing provisions can be avoided.

ECONOMIC AND FINANCIAL APPRAISAL

Economic Analysis

The purpose of an economic appraisal is to demonstrate that the proposed development is of sufficient economic merit to justify an investment in it. The appraisal compares the costs incurred for the whole scheme over its lifetime—capital and operating costs—with those of an alternative solution that can meet the same objectives. This means that the two projects to be compared must be able to supply the same amounts of energy, at the time required, to satisfy the same demand. Any restrictions imposed by this demand must be taken into account in deciding on the output with which the projects can be credited. The reason for this kind of comparison is that the power market that the proposed project is to supply must be assumed to require a specific amount of electricity from one source or other. The problem is to find the cheapest source. The economic viability of this source is established in terms of the savings or extra benefits that it achieves in comparison with the alternative solution.

The alternative solution for purposes of economic comparison with a small-scale hydro scheme is usually:

In the case of a supply given to an isolated load center: a diesel plant.
In the case of a supply to be given to the existing network: either an alternative

diesel plant feeding into this network, or an extension of existing generating capacity elsewhere in the network, in which case the average production cost of the power supplied to the network is the relevant figure for comparison with the production cost of the hydro scheme. If a network extension to a new load center comes into question, instead of the installation of the hydro plant, the economic comparison is again made on the basis of the appropriate average production costs in the two cases. "Average" is the mean annual production cost per kW delivered by the power plants to the main power infeed points in the network. Transmission costs and losses are included only for the lines between the power plants and these infeed points. For an isolated load, the relevant infeed point is where the power is transferred to the local distribution system.

Electricity tariffs or the revenues obtained from them are not recommended for use in the economic appraisal of a small-scale hydro scheme. Computation of revenues from the sale of the hydro energy to the final consumers or from the sale price of a network supply can be a very complex matter because several different tariffs and consumer groups can be involved. An average network tariff rate will be valid only for a specific load pattern and constellation of consumers; it will change with changes in the power demand and in the composition of the consumer groups. Revenue computations are appropriate only for the valuation of a speculative demand, for example, for the supply of an isolated power-intensive industrial project, which will be built, jointly with a dedicated hydro-plant, only if the power costs do not exceed a given ceiling.

The economic appraisal of alternative solutions relies essentially on comparison of the present-valued cash flows for each of the alternative solutions. The case flows comprise:

Capital costs of each power scheme and its associated transmission, phased in accordance with the annual incidence of expenditure during the construction period and including both the initial installation and any plant replacement during the period over which the cash flow is computed.

Annual operating, maintenance, administrative, and insurance expenses directly attributable to each scheme; expenses common to both solutions are not taken into account.

Annual fuel costs of thermal plant entering into the comparison.

The costs of a thermal back-up plant required to firm up the output from the hydro sheme during dry periods have to form part of the cash flow for this scheme; these costs will also include a fuel component.

The present-worths are established by discounting the cash flows arising throughout the lifetime of the longest-lived asset in the comparison—generally the civil works of the hydro scheme—to the date on which construction began. This date is taken to be the "inception date" for the analysis. Construction of the thermal alternative can be accomplished more quickly and its cash flow can start a year to two later, but the plant must be ready for operation at the same time as the hydro scheme.

Project components having a shorter life than the longest-lived asset—hydro and thermal generating plant, electrical equipment, and transmission—will need to be replaced at the end of their particular lifecycle.

Expenditure incurred prior to the inception date is charged with interest at the same rate as the discount rate adopted for present-valuing the cash flows; this antidated expenditure and the interest on it are then added to the present-worth to make up the total project costs. Preinvestment expenditure should be taken into account only if it has in fact been paid for by the developer out of loans or borrowings; grant aid should not be included. Capital expenditure incurred during construction is not charged with interest when it is discounted.

Calculations are normally done at constant price levels; it seems hazardous to speculate what price escalation might be appropriate for the longer term future. The potential impact of price escalation can be tested by sensitivity studies.

The principal parameters for the present-valuing process are as follows:

1. Asset lives. Asset lives used for economic calculation denote a notional period related to the length of service that can safely be expected from a new installation. The life extends up to the time when the reliability of the scheme seriously deteriorates and repair and replacement of the equipment becomes a major expenditure item. Asset lives are often set by the financing agencies involved in the project. Lives lie generally between 50 and 70 years for the civil works and 25 and 35 years for electro-mechanical equipment for both hydro and thermal plants. Diesel engines are given a life of 15–20 years. Some agencies require a more detailed breakdown of lives for individual project components, but this makes calculations more difficult and is usually not worth doing for a small scheme.

   Computation is greatly simplified if the asset lives are chosen in multiples, say 15 years for diesel units, 30 years for electro-mechanical equipment, and 60 years for civil works. There will then be two replacement cycles for the diesel plant and one for the hydro equipment during the 60-year lifetime for the scheme as a whole over which the economic evaluation will extend. A longer asset life will have very little effect on the present-value because the discount factors become very small in later years.

   The physical life of the scheme depends greatly on local conditions and may well differ from the notional asset life in specific cases. The financial life of the scheme is governed by the loan conditions—the loan life—under which it is financed. Loan lives are generally taken to be about 10–15 years (including a moratorium on repayments during construction) for private-sector financing and 20–25 years for financing from public-sector institutions. A notational loan life should also be assigned to finance obtained on grant aid so that the output from the scheme can be realistically valued.

2. Discount rates. The discount rate for economic analysis should be pitched close to the average rate at which capital can be borrowed for public sector investment. If the discount rate is set too high, the benefits from the scheme will

become undervalued; if the rate is too low, the benefits will be exaggerated. A usual figure is 10%, with adjustment within a range of +2% where justified by borrowing conditions or fiscal policy. The effect of different discount rates can be explored by sensitivity tests.

Several methods of economic appraisal are commonly used; they all involve comparison of two alternative solutions, but they differ in detail.

1. Benefit/cost analysis. The benefit of the project is the saving it achieves over its lifetime in comparison with the alternative solution. The benefit is computed from the difference between the present worths of the hydro scheme and its thermal alternative. The cost is given by the present-worth of the cash flow for the hydro scheme. The cash flows used for the analysis are always discounted at a fixed rate. If the hydro scheme is to be economically viable, the benefit must exceed its cost by a margin no lower than the discount rate used in the computation, although some authorities accept lower rates of benefit on social grounds.

   Another method is sometimes appropriate when a new scheme supplies an existing network or is to be compared with an alternative network supply. In this case, both benefit and cost are derived from the respective production costs of the energy supplied.

2. Internal rate of return. The cash flows of the alternatives are here present-valued at different discount rates. The internal rate of return denotes the rate at which the present-worths of the two cash streams are equal. This rate shows the return to be expected on the project that has the higher initial investment cost, usually the hydro scheme. The break-even rate is used as an index of economic merit. The merit is considered adequate if this rate is close to or above, the opportunity cost of capital in the economy in question, or at least equal to the return that might be achieved from public sector investment in general.

3. Comparison of production costs. The production costs are made up of capital charges (interest and depreciation for the whole scheme, including transmission), operating and maintenance costs, fuel costs, and project-related administrative expenses. These costs are computed for the amount of energy sent out to the load centers and they include transmission losses up to the point of delivery of the power. The amounts of energy and some of the operational costs are likely to vary from year to year during the lifetime of the scheme and it is therefore necessary to present-value the annual streams of costs and energy. This process is different from that previously described for two reasons:
   - The project has to be depreciated over its financial life, not its asset life. This means that the cost and energy streams need to extend only over, say, 20–25 years.
   - Interest at the real borrowing rate is payable on the capital expenditure as soon as it is incurred. Interest must therefore be charged on the phased

capital expenditure during construction, but there will probably be no operational expenditure during the construction period. A conventional point for present-valuing is the time of commissioning when the project begins to be depreciated. Annual sums for interest and depreciation—the latter usually on the sinking fund method—are now included in the cost streams (with 10% interest and a 25-year loan life, capital charges amount to 11.017% per annum).

The ratio of the present-worths of the two data streams—costs and energy—will present the "normalized" unit production cost, per kWh delivered, appropriate to the particular solution.

If it can be assumed that operational costs and energy quantities sent to the power market will remain the same in every year, the computation can be simplified by determining:

- the annual charges on the total capital expenditure (with interest during construction) at the time of commissioning
- the annual operational expenditure (including maintenance, administration, fuel costs, and losses)
- the annual amount of energy delivered (which can be an average computed over a longer period)

The production cost per kWh can be computed directly from these figures, and present-valuing will not be necessary.

The production cost can be an important parameter for optimizing the design of the hydro scheme. Where changes in design can affect the amount of energy produced, the least cost solution is found by assessing the extra cost of a given design change and determining its effect on the energy cost. Costs common to all solutions can be left out of account.

4. Comparison of present-worths. This method is sometimes used as an index of merit, but it can give misleading answers if the outputs of the two schemes are not precisely the same in every year of service. It is not recommended.

Sensitivity studies are needed to explore the effect of changes in the parameters entering into the economic appraisal. Such studies are especially useful for investigating borderline situations and helping in coming to a decision on projects of questionable merit where social and policy considerations may come into play. Shadow pricing can also be introduced to deal with solutions that rely on a substantial import component. Taxation is generally not taken into account.

Financial Analysis

Having shown that a given solution is of adequate merit, it is usually necessary to present a financial analysis of the proposed scheme to the potential financing agency. Most agencies will prescribe what information should be provided; it can range from a simple financing plan showing phased borrowing requirements to a complete set of proforma accounts for the utility concerned, with the new plant framework for construction and operation, the scope of which will depend greatly on the institutional arrangements under which the plant is to operate.

In its simplest form, the financing plan requires no more than the setting down of borrowing requirements to meet the anticipated flow of construction expenditure. Particular items of expenditure may have to be allocated to specific sources of finance, especially if any tied or aid-supported finance is involved, as is often the case in developing countries. A financing plan can then be drawn up in which sources and requirements of funds are properly matched and the overall loan commitments and their implications clearly demonstrated. The developer must be aware of these commitments and must be satisfied that the new facility will provide sufficient revenue to meet these commitments, which may be in both foreign and local currencies. The financial analysis should demonstrate what the commitments are likely to be.

More comprehensive accounts involving statements of assets and liabilities and of sources and applications of funds are to demonstrate to the lending agencies to what extent the developer can incur additional commitments for the proposed scheme, bearing in mind the standing commitments which the developer has incurred already, and to what extent the developer—as opposed to the scheme—is creditworthy. Although loans are usually forthcoming if a scheme is of sufficient merit, the lending agency may adjust the loan conditions to the financial situation that emerges from the analysis if it may impose preconditions, such as the raising of tariff levels, to achieve a better return on the assets in service. Where the financing package includes an element of private sector funds or suppliers' credit, the financial soundness of the borrower may have to be demonstrated before any contracts are let, although financial risk insurance is now widely used—especially for exports—to facilitate private sector participation in development projects.

FINANCING ARRANGEMENTS

Financing a small-scale hydropower project poses several problems for developing countries, brought about primarily by:

The relatively large capital requirement, per kW installed, for both the preinvestment and investment phases of the project

The small overall size of the total financing package in comparison with the cost of power sector development in general, and the large proportion of its foreign currency component

The extensive administrative effort needed to launch the project, harness local resources for it, and attract foreign support

These matters require consideration throughout the development process and need to be finally resolved when the feasibility of the project is established and a financing plan has to be laid down by the developer.

A financing problem first arises when a project is conceived and studied. Although the technology involved in small-scale hydropower is well explored and fully proven, the technical resources available to developing countries are often not adequate for initiating the preinvestment work and for carrying it through without foreign technological support and a consequent foreign input. This input is more speculative in the early stages of a project when its chance for success will not yet be known. The problem of securing foreign support is more readily resolved if the developer has adequate funds at his disposal for recruiting whatever assistance may be needed. If funds are not available, the developer will have to enlist assistance from multilateral or bilateral agencies or private organizations.

Multilateral agencies comprise intergovernmental institutions such as the United Nations system, the World and Regional Development Banks, and multinational funds especially set up by groups of countries in Europe, the Middle East, and elsewhere, all charged with the task of supporting social and economic progress in the developing world. Multilateral agencies are generally free of national bias and will facilitate access to international markets for services and equipment.

Bilateral agencies are national and governmental bodies set up by some 20 industrialized countries—sometimes termed "donor countries"—for the same purpose as the multilaterals but influenced to a greater extent by national policies toward development assistance. Bilateral agencies generally limit the support they provide to the use of nationally tied suppliers of services and goods.

Private organizations providing technical assistance can vary in form from institutions to corporate bodies and generally offer services and goods only from their own sources. The term "generally" implies that there are exceptions because:

Some multilateral agencies embrace a limited group of countries and restrict financial support to supplies of services and goods from these countries.
Some bilateral agencies do not always insist on tied supplies.

The scope available under given financing conditions has to be explored before financing arrangements are concluded so that the developer is quite clear on the terms he is committed to and on the merit of these terms.

Funds for preinvestment expenditure are fairly readily found because the sums involved are not large in absolute terms, the socioeconomic impact of the work is considerable, and technology transfer is most effective at that stage. There can be no installation on the ground unless the preinvestment phase has been satisfactorily completed. Technical assistance on grant aid or favorable terms is therefore available from several sources, limited to the study sector or combined with a financing package for the scheme as a whole on the assumption that the studies will be successful.

In all cases, the developer will need to bring judgment to bear to ensure that the arrangements offered do not lead to commitments with which he will not be able to comply in the long run; the developer should seek impartial advice to help him deal with conflicting issues.

A major difficulty often is the funding of the local currency component of the project, which may amount to 40–60% of the total investment and usually covers most of the costs of the civil works. Although maximum local participation is important from the social and economic aspect and can greatly reduce foreign currency requirements, securing adequate provision in local currency can be difficult. Where the developer forms part of the national power utility, or a branch of it, local expenditure can probably be paid for out of the budget established for power system expansion or dedicated to rural electrification. Where the developer forms an isolated entity, say a rural cooperative, funds might be made available from a rural development budget set up by central or local government. A contribution in the form of local labor and materials from nonmonetary sources might also be secured in appropriate cases, compensated for in kind—through supplies of foodstuffs, fertilizer, housing, and ultimately electricity. Foreign funding support is not normally available for local expenditure, but support in kind can sometimes be obtained in cases of special need.

The financing of the foreign expenditure component of the investment phase covers a wide erange of possibilities. Much emphasis is now placed on innovative financing, which includes different terms in the same package. The most common types of financing are:

1. An outright grant for the whole or part of the project; 100% grant aid is usually given only for special or demonstration projects that aim to introduce the concept of small-scale hydropower into the energy supply scenario of least-developed countries and can offer a basis for a more extensive hydropower program. Partial grants are more freely given, especially in combination with other financing arrangements.
2. Soft loans carrying no or very low rates of interest (1/2–3% p.a.), extending over a long period of time (25–50 years), and including a moratorium covering both the construction period and the early years of operation. As in the case of grants, soft loans can be mixed in with other terms in an overall package.
3. Loans at concessionry interest rates, 1.1/2–8% p.a. below commercial rates, and provided by national credit organizations either directly or through suppliers of services and goods. Some multinational groups, such as the Organization for Economic Co-operation and Development (OECD), have established "consensus rates" in the range of 6–9% p.a., which their members can apply to loans for development projects. The loan life is usually of the order of 15 years, but can be both shorter and longer in specific cases.
4. Loans from multilateral financing agencies, principally the development banks, which carry an interest rate of 8–10%, a loan life of about 25 years, and a moratorium on repayments during the construction period. Inclusion of a soft loan component can greatly improve the attractiveness of such loans.
5. Loans provided by commerical banks on commercial terms with interest rates at normal bank lending levels and a short loan life, often less than 10 years.
6. Buyer's credit with credit facilities on consensus or commercial terms arranged for the developer but usually restricted to procurement from a limited range of national suppliers.

7. Supplier's credit under which services and goods are provided on credit terms linked to the best loan conditions that the supplier can negotiate. The terms generally lie between consensus and commercial rates but suppliers sometimes subsidize the loans through commercial rebates on their supply. An element of grant aid can also be included.

8. Mixed credit and cofinancing terms are offered in conjunction with turnkey or package deals, an arrangement that is particularly relevant for small hydro schemes where the size of the foreign investment component does not justify placing separate contracts for individual plant items. A typical package will include some aid or soft loans and consensus conditions for supplier's credit. The package will be tied to prearranged suppliers (selected by the tenderer, not the developer), but it is often financially attractive.

Grants and soft loans clearly present the most favorable financing conditions for a complete project, but they are available only in special circumstances where there is a strong social and economic need. The average developer will have to choose between the various loan terms on offer and will probably obtain the best conditions from a package deal. The developer will be well advised to obtain competitive offers for an inclusive package, which might also include a feasibility review and detailed design.

Development banks have the advantage of contractual impartiality. They offer the developer technical and administrative support that is particularly valuable if he is less experienced. They do not restrict the choice of suppliers and provide guidance for selecting the best deal. Their loan terms can be very favorable if an element of grants or soft loans is included, but failing this, their loan conditions are often not competitive with supplier's credit or package deals supported by national export promotion schemes. Many development banks also find the small size of the financing requirements in foreign currency—in the range of US$1–2 million for a 1 MW scheme—unattractive. They usually demand a bankable feasibility report before they can consider to what extent the project is creditworthy and this can add to the cost of the preparatory work unless it can be covered under an aid program.

Buyer's credit leaves the initiative for selecting suppliers and contractors with the developer, which can permit competitive tendering, not only for individual plant items but also for complete packages of the electro-mechanical equipment. Where credit is nationally restricted, the choice of potential suppliers may, however, be so limited that no advantage can be derived from competition. Although buyer's credit may not be an attractive option in these circumstances, it should be looked into for comparison since the alternative—supplier's credit—likewise involves no competition. Care needs to be taken in a noncompetitive situation to ensure that the plant and equipment offered are suitable for the duty they have to perform.

Loans on purely commercial terms have little appeal for a small development project. Negotiations for financing arrangements can be complex and time-consuming, even where aid programs are involved. Negotiations should therefore begin as soon as project parameters have been identified and the developer has serious intent to proceed with the project.

10

Development

PROJECT PARAMETERS

Hydro power is critically dependent on the physical parameters at a specific site; these parameters must therefore be ascertained as reliably as possible before a successful scheme can be developed. The problem often faced with a small-scale project to be built in an outlying and only cursorily explored area is the lack of reliable basic data. This problem must be clearly recognized and dealt with by thorough site surveys and correlation studies with neighboring areas showing similar characteristics that may help to confirm the available site information. A further problem is that in order to keep preinvestment expenditure in check, the lead time and the extent of the site investigations have to be restricted but without impairing the reliability of the project. A balance has thus to be struck between the scale of the investigations and the dependability of the results; dependability is crucial for success.

The principal parameters to be established before any design work can be put in hand are:

water availability (hydrology)
surface conditions of the ground (topography)
subsurface conditions (geology)
ground stability (seismicity)
water quality (siltation and sedimentation)
environmental aspects (ecology and morphology)

In addition, the size and location of the power market, the accessibility of the site, and its security against physical disturbance all have to be ascertained.

Electrical output

The electrical output from the scheme is governed by the water runoff or flow through the turbine (Q) and the effective head of water that can be developed above the turbine (He, which is equal to the total or gross head less friction losses in the pressure conduits). The power produced at the generator terminals is linked to these two quantities by the equation:

$$P = 9.81 \times Q \times He \times e$$

(where Q is in cubic meters/second, He is in meters, P is in kW, and e is the product of the turbine efficiency (83–92%) and the generator efficiency (93–96%). For a small-scale (1–5 MW) and medium-head (150–250 m) power plant using the horizontal-shaft Francis turbine, $P = 8.2 Q \times He$ is a rough approximation.

The power delivered to the load center will be reduced by the in-house electricity requirements of the power station—for lighting and control equipment—and by transformer and transmission losses.

The gross head is fixed by the local topography and cannot be changed once the scheme is fitted into the topographical features of a given site. The water runoff, on the other hand, can vary within wide limits in a generally uncontrollable, and indeed often unpredictable, way. Careful exploration of the hydrology is therefore an essential precondition for the realization of a hydro space.

Hydrology

The hydrology of the river system on which the hydro scheme is to be based is dependent on the climatic and rainfall pattern to which the particular water catchment area is subjected, but not exclusively so; water transfer from other catchment areas— naturally or by design—is also possible. The hydrological conditions can vary both cyclically—between the wet and dry seasons—and at random from year to year with changes in climate and rainfall. Some measure of the cyclical changes between flood and drought can be obtained from observations extending over only a few years, but determination of the multiannual runoff characteristic requires observation of water flow in the river system over a long period of time—at least 30 years for a major scheme. Such information is not usually available for a small scheme in a rural or remote area, nor is it possible to defer development until adequate hydrological information has been collected. On the other hand there is danger of:

Overdesign for runoff conditions that are not achieved in practice, or
Underdesign if the size of the flood flows has not been adequately recognized

Inadequate exploration of the hydrology therefore entails a risk. The risk can be reduced but not entirely removed by using various methods of correlation analysis in cases where local hydrological data are inadequate. The principal methods used are:

1. The development of runoff data from measurements of the rainfall pattern in the catchment area or in an area with analogous characteristics.

2. Correlation of runoff conditions with those of nearby river courses for which more extensive hydrological observations are available.
3. Computerized hydrological models that relate rainfall data to subsoil conditions in the catchment area and make allowance for water losses and transfer to the river system for which the runoff pattern is then derived. The reliability of this method can be improved by correlation with actual flow measurements in the river concerned or in a neighboring water course with similar characteristics.

In practice, all three methods will probably be called upon to firm up an inadequate series of measurements of river flow that is to be diverted through the hydro plant. It is advisable to initiate flow measurements as soon as the hydropower potential of the river course is being examined for exploitation and to continue these measurements at least until the design of the hydro plant is finalized. Further measurements will be of help if other river courses in the vicinity are to be exploited at a later stage and data for correlation analysis are to be assembled.

Flow-duration curves can be plotted from the observed—or computed—relationship between the water flow through the power plant and the period of time in the year for which this flow is obtained. The duration curve usually denotes the amplitude of the flow, in cubic meters per second, in terms of the percentage of time in the year during which this amplitude is exceeded. A flow duration of 93–97% of the time (8,147–8,497 hours per year) is termed "firm" and the corresponding water flow is considered to be available at all times. Flow of lower duration is "nonfirm" or seasonal.

If the flow quantities are multiplied by a constant for the particular scheme, say $8.2 \times Q \times H$ in the example quoted earlier, the flow duration curve is converted into a power duration curve, which indicates the amount of power that can be generated for given fractions of the year.

In the power duration curve shown in Figure 10.1, firm power is assumed to be generated for 95% of the time (8,322 hours per year); the equivalent firm power capacity of the plant, that is, with the plant run continuously, is then given by the duration curve. Seasonal power is produced in this example for $95 - 22 = 73\%$ of the time (6,395 hours per year). The maximum capacity installed in the plant uses a flow Q for 22% of the time, but this output is not firm. Any power demand between Qmax (95%) and Qs cannot be met at all times by the hydro plant and thermal power compensation may be needed. Flood flows in excess of Qs are not used for power generation and are spilled.

Two quantities thus determine the design of the power plant:

maximum firm output at Qmax (95%)
maximum installed capacity at Qs

The value of Qs and the corresponding capacity to be installed in the scheme are chosen on a basis of cost, which includes the civil works as well as the power plant

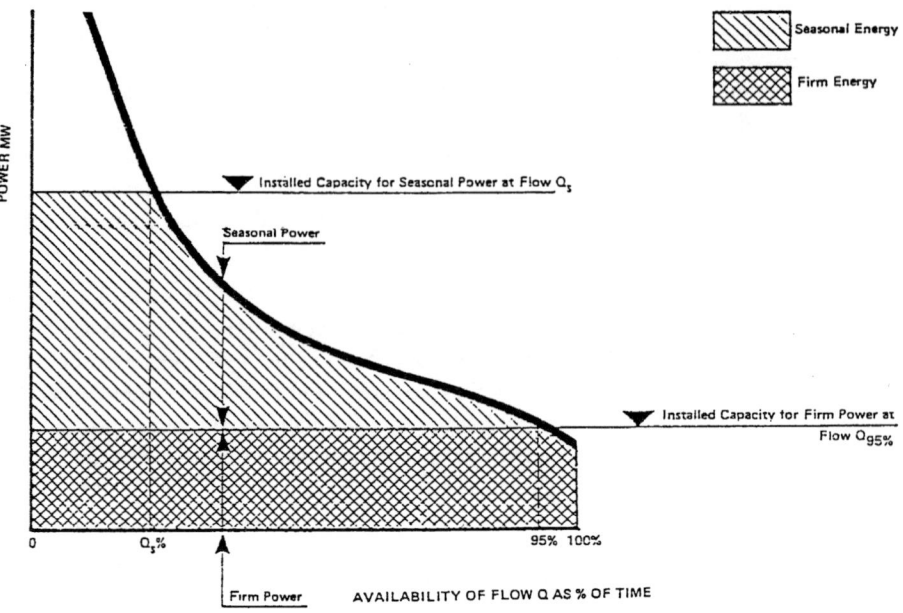

Figure 10.1 Power duration curve.

cost, and power demand. If there is sufficient power demand, the installed capacity needs to be optimized for least cost of the total energy produced.

If water storage or pondage can be incorporated in the scheme, the firm energy component will be increased, depending on the proportion of the flow that storage permits to retain and regulate. Adaptation of the nonfirm component to the load pattern will also be improved.

Topography

A topographic survey of the project area has to be undertaken to establish two essential features of the potential scheme:

The head of water that can be developed between the intake to the headrace or pressure conduit and the power plant

The surface characteristics of the project area that govern the design and layout of the scheme and largely determine its feasibility and cost

The local topography must permit diversion of the river flow through the power station. It must enable the distance between the intake and the power plant to be kept reasonably short to minimize hydraulic losses in the headrace and pressure conduit system and also to minimize the construction costs of this system. It must provide reasonable space for the power house and access to the scheme for construction, operation, and maintenance and also for the transmission link to the load center. It must allow the water discharged from the plant to be returned to the river. It must make it possible for flood flows to be spilled without endangering the civil works and the power station.

A first impression of the local topography can usually be gained from a study of maps, and the most promising area for developing the project can thus be identified. This has to be followed by a detailed survey on the ground, which is to establish the river profile, the extent of the catchment area, the location of the diversion dam and intake, and, in suitable cases, of the headpond and reservoir. The survey must furthermore establish the precise routing of the headrace and pressure conduit and the surface profile and condition along this route from which will have to be concluded whether it is feasible to run the conduit system on the surface or whether undergrounding will be necessary. The survey will also yield information on the siting of the power house, tailrace and spillway, and the alignment of the transmission line. Maps to a scale of 1:1000 with contour intervals of at least one meter have to be prepared for the precise siting of all the structures. Detailed cross sections may have to be drawn up at critical points along the surface profile. Subsoil conditions will also play a role in the decision on the precise siting of these structures. Correlation between surface and subsurface investigations will therefore be necessary.

Geology

The ground throughout the project area must be load-bearing and stable and the water table must be low enough not to affect the proposed structures. A first approach to geological mapping of the project area is to expand surface mapping by visual observation and sampling of subsoil conditions, followed by laboratory analysis of the samples obtained. The sampling campaign will have to be more searching if there are indications of a complex geology, in particular:

- major faults or unstable rock in or close to the project area
- heavy weathering of the subsurface strata
- indications of ground permeability or fissures in volcanic areas or limestone
- instability of the ground

Detailed geological investigations can be expensive and expert advice should be sought before they are undertaken. Even though the demands on the foundation conditions may not be very great for a small scheme, the geological information available for a remotely located project area may well be sparse, and construction may then entail some risk. The risk will be greater if tunnelling is needed for water adduction or pressure conduits, or even for underground location of the power station in some cases; a thorough subsoil survey will then become essential.

Geological exploration involves boring in the first instance, which can also help to determine the extent of the overburden. Test pits may be needed to examine specific foundation conditions and trenching may have to be used if the bedrock is to be exposed for the siting of major structures, especially dams and diversion weirs. Exploration by core drilling, tunnelling, or addits provides reliable information but is usually too expensive for a small scheme.

A geotechnical soil survey may well be required in cases where inadequate ground stability is suspected or where the water table is unduly high and land is liable to flooding. Ground stability can be affected by the construction of reservoirs and even headponds. Design parameters for a small scheme can, however, often be derived by judgment and costly geotechnical investigations reduced to a minimum.

The geological survey is also to yield information on the availability and suitability of local materials for fill and concrete aggregates, on the volumes to be excavated, on the handling and disposal of the excavated material, and on construction costs.

Silting and Sedimentation

The carryover of solids in the river water can cause damage to the structures and turbine blades through abrasion and erosion; it can also cause settling-out of sediment in places where the water flow is low, especially in reservoirs and stilling basins. The sediment load in the river generally increases with the amount of water carried and is greatest at times of flood flow.

The solids content in the river water is determined by sampling both suspended and bed load sediment at regular intervals and at different flow conditions throughout the year. Sampling must be coordinated with flow measurement so that the two quantities—solids content (weight of volume of sediment per unit time) and rate flow—can be properly related. The size distribution and composition of the sediment should also be analyzed. This information, supported by a theoretical study of sediment entrainment and transport in the water, is needed for the design of components exposed to the water flow so that erosion damage can be minimized.

The overall design of the scheme will have to take account of the potential sediment load and incorporate by-pass and desilting arrangements, sand excluders, and sand traps, to limit the carryover of silt into the waterways and protect the scheme against damage and deterioration. Periodic purging of silt-retaining features and inspection for erosion damage will be important maintenance tasks.

Seismic Considerations

Where seismic activity could affect the stability of the structures, investigations into the nature and frequency of seismic disturbance and the magnitude of the seismic stresses will be necessary. A full range of seismic studies cannot normally be justified for a small scheme and data already available for the project area are therefore used to provide a basis for the seismic design of the scheme. If the available data are inadequate of if their reliability is in doubt, a decision needs to be taken on the need for complementary studies and a justification for them. If, on the other hand, the uncertainties and risks appear paramount, the particular site under investigation may have to be abandoned.

Ecology and Environment

Any change in ambient conditions that the construction of the hydro scheme may bring about will have an impact on the local ecology and environment and will have to be taken into account in the design of the scheme. An appraisal of this impact will therefore have to be made when the design parameters are being determined. The principal factors to be looked into are:

1. The effect on the ecology resulting from changes in water quality in the river, between the points of water intake and discharge, due to the water abstraction and in reservoirs due to stagnant water; both may lead to changes in the morphology.
2. Degradation of the riverbed downstream due to sediment retention by the scheme and a consequently increased sediment pick-up capacity of the water discharged from the plant.
3. Changes in land use, and the consequent impact on the local economy resulting from: changes in the water table due to water abstraction and storage; prevention of other uses for the land occupied by the scheme; reduction of soil fertility in the river basin downstream because of silt retention by the scheme.
4. Visual and noise impact during both construction and operation.

Remoteness of the site should not discourage careful appraisal of environmental consequences because the hydro scheme is to become a focal point for local development and is to remain in existence for a long period of time.

PROJECT COMPONENTS

General Concept

Selection, layout, and design of the project components making up the hydro scheme depend on the site parameters as previously determined. The principal design criteria are the following:

For the dimensioning of the project components: the head and water throughput of the scheme.

For the design of the scheme (intake works, headrace, and pressure conduits, power plant and tailrace): the site conditions.

For the rating of the power plant and the transmission system: the size of the power market and its location.

Every scheme has to meet strictly local conditions and must be designed specifically for these conditions. Standardization of design can therefore be practiced only to a limited extent, although some uniformity of approach should be introduced where ambient conditions are similar. The design characteristics of small hydro schemes cover a wide range of possibilities and can be described only in general terms.

Scheme Layout

Typical layouts for small schemes of medium head are illustrated in Figure 10.2. Schemes of this type cover the majority of cases. The river is diverted by means of

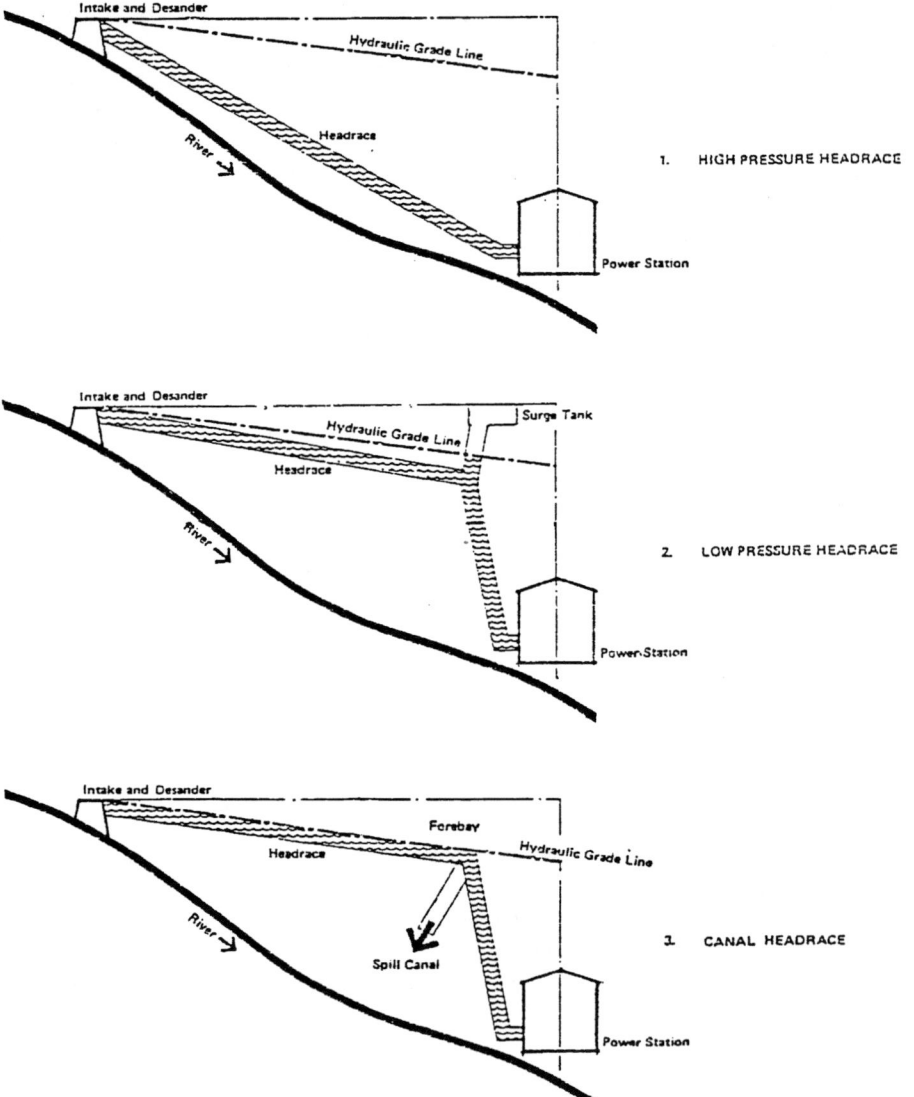

Figure 10.2 Scheme layouts.

a weir and the water is conducted to the power station in various ways depending on the contours of the ground and on the length of the headrace (L) in relation to the head (H). A reasonably steady slope between intake and powerhouse and an L/ H ratio of up to about 5 will permit use of a single high-pressure conduit or penstock (case 1.). In many cases, the head can be concentrated into a steep drop close to the power plant and the headrace then divided into a low-pressure and a high-pressure section (case 2). Where the pressure drop in the headrace is sufficiently low and the contours are suitable, the headrace can be in the form of an open canal; this is often the cheapest solution (case 3). A water flow rate of at least 1 meter/second must be maintained in the headrace to avoid sedimentation. The hydraulic gradient line in the diagrams of Fig. 10.2 indicates the minimum slope needed to ensure least flow conditions and permit use of an unpressurized headrace.

Intake Works

The diversion weir has to be built at a narrow section of the river and founded on firm rock. A more substantial structure—a dam—will not generally be justified for a small scheme, especially as the flow rates required at medium head and hence the amounts of water to be diverted will not be large. Where there is danger of very coarse material—rocks and gravel—being carried in the river, a pit is excavated upstream of the weir to trap this material. The design of the intake works has to take account of the velocity of the river water and provide for:

The by-passing of flood flows, usually by arranging for overflowing the weir and allowing it to act as a spillway.

The trapping of silt and sediment in a sand trap or intake pond with facilities for periodic or continuous flushing out of the deposited material (continuous flushing requires an adequate water supply).

The retention of large trash by means of a trash rack or screen at the entry to the headrace; regular trash removal is essential to avoid excessive head loss from clogged screens.

A case for a storage reservoir at the intake can be made for only very small schemes, but provision of pondage will permit some output regulation under low flow conditions and will be of advantage. Pondage can sometimes be combined with a stilling or settling basin serving also for sediment retention and silt exclusion from the waterways. Where an intake canal can be built or combined with the headrace canal, the stilling basin can be located further downstream if the ground conditions permit, but arrangements for spilling and purging will have to be incorporated, which can make this solution more expensive (see Fig. 10.2, case 3).

Headrace and Pressure Conduit

Economy requires the water conduit system to be located on the surface, as far as this is possible, but covered for protection against damage. The system can be in the form of a single penstock or divided into low-pressure and high-pressure sections as already described. In the latter case, a surge tank has to be provided at the joint between the two sections to absorb the hydraulic shock arising from operation of the turbine inlet valve and to avoid water hammer. A surge tank is essential when the L/H ratio exceeds 5, but it is not needed if the inlet to the high-pressure conduit is open to atmosphere or replaced by a forebay (see also Fig. 10.2, cases 2 and 3). The economic limit for the length of the headrace for a small plant lies at an L/H ratio of about 30.

The cross section of the pressure conduit has to be optimized for least losses and least cost under the rated flow conditions. The conduit should be able to withstand a water pressure about one-third higher than the gross head; this pressure is broadly equivalent to the maximum permissible water hammer stress. The design of anchorages for the penstock and the cost of laying it depend largely on the nature of the ground.

Steel is the most versatile material for the penstock and can be used for any length and head. It is expensive to transport to site and manipulate and needs to be protected against corrosion before being buried. Expansion joints have to be provided, especially for exposed sections of the pipeline. Malleable cast iron offers an alternative of about equal weight and strength but better resistance against corrosion. Plastics are suitable for heads of up to 150 meters. The most interesting material is glassfiber-reinforced epoxy resin; it is light and easy to transport—with a specific weight of about one-fifth of that of steel and iron—but it does not have the same strength and needs to be protected against local stresses and impact. Its expansion coefficient is about twice that of steel and this has to be compensated for by expansion joints. Other plastic materials, PVC and Polythene, are also light in weight and lack strength; they are of only limited applicability. Concrete pipes are durable but expensive. Wood is suitable only for short runs and has a short life.

Headrace and pressure conduits have to be regularly inspected and access along their route must therefore be available. Means of draining possible seepage will also have to be provided. Gates, usually of the sliding type, located at the head of the penstock will permit closure of the waterway for maintenance and in an emergency, but they are slow acting and may have to be backed up by an emergency shut-off valve to avoid damage to the plant.

Powerhouse

The powerhouse contains the turbo-generating plant, the associated valve gear, and the control and supervisory equipment for the plant. The size of the power plant depends on the head and rated water throughput at full load. The dimensions of the powerhouse must allow for the number of units of this size to be installed and for the space required for erection and maintenance of the plant. An annex is often built to house the transformers. The height of the powerhouse must be sufficient to permit the removal of the turbo-generators by means of an overhead travelling crane, which should have a load-bearing capacity at least equal to the weight of the generator rotor. Load-bearing walls are required for support of the travelling crane. The powerhouse foundations must be strong enough to absorb the vibrations from the rotating machinery and to withstand the full penstock pressure when all the turbine inlet valves are closed. External treatment of the powerhouse structure may be required to reduce visual impact.

The two types of turbine inlet value most frequently used for small units are:

Spherical or ball valves for heads in excess of 150 meters
Butterfly valves for heads below 150 meters

Both types are normally designed to withstand a 50% excess over the combined static head pressure and dynamic surge pressure arising during valve closure at maximum flow. Sluice or gate valves are suitable for high head pressures, but they are expensive and are now used only with penstocks of up to 0.5 meters in diameter. All valves are designed to close by gravity to permit emergency shut-off.

Provision has to be made for a supply of cooling water for the rotating plant, generally by pumping from the tailrace. Draining arrangements are also necessary to protect the powerhouse against flooding.

Tailrace

Water leaving the hydro turbines is discharged into the tailrace. Its function is to maintain the water level at the turbine outlet at a point appropriate to the required setting of the machine so that cavitation of the runner is avoided. The tailrace must also protect the power plant against reserve flow from the river under flood conditions. This is usually achieved by a weir at the outlet of the plant; the crest level of the weir is at the anticipated maximum flood level of the river. Precautions have to be taken against erosion damage caused by high water discharge velocities over the weir into the tailrace channel leading to the river.

Electrical Equipment

Equipment has to be installed in the power station for:

Control of the turbine generator speed, machine loading and start-up sequence, and for control of the switchgear.
Monitoring of speed, water flow and level, temperatures, and electrical parameters at low voltage and high voltage (voltage, current, frequency, power factor and output).
Protection of the generator and transformer against electrical faults and of the rotating plant against overspeed.

A transformer steps up the generator voltage to the transmission level. The transformer can be connected directly to the generator busbars if it is supplied by only one machine; otherwise low-voltage switchgear will be interposed between the bus-bars and the transformers. High-voltage switchgear will be installed between the transformers and the transmission line. The switchgear can be of either the indoor or outdoor type. An outdoor switchyard should be located close to the powerhouse and protected against flooding.

TURBINE CHARACTERISTICS

The hydro turbine converts the kinetic energy contained in the water runoff into mechanical energy, which is used in turn to drive the generator producing the electrical output. Choice of the turbine depends primarily on the available head and on the water throughput; but it is also influenced by the cost of the machine, its efficiency, its part-load characteristics, and its setting with respect to the configuration of tailrace and power station civil works. Turbines are generally classified under:

- impulse machines, for high and medium head applications
- reaction machines, for medium and low head applications

In the *impulse turbine*, a free jet of water impinges on the turbine wheel or runner and is then discharged into a pit below the turbine before passing out of the station through the tailrace. Load variation is achieved by controlling the force and direction of the incoming jet of water. The most usual types of impulse turbine are:

Pelton turbines, in which the water jet strikes a series of buckets arranged along the periphery of the turbine runner.
These machines are generally suitable for heads from about 1,000 meters down to some 150 meters. The runner shaft is usually arranged horizontally. The turbines can be fitted with a single jet, which provides a simple and space-saving arrangement and is used for small outputs at high heads. With two or more jets, a still more compact arrangement is possible because a higher turbine and generator speed can be achieved and the runner diameter reduced to make the unit more cost-effective. The output range of the machine is then greater since only one jet needs to be used. Pelton turbines are not normally subject to cavitation and can be controlled down to virtually no load without difficulty. Their part-load efficiency is also generally higher than that of reaction turbines, but the efficiency falls off slightly above about 70% full load; the efficiency reaches about 90–92% between half and full load.
Turgo impulse turbines, a variation of the Pelton type. The incoming water jet impinges on the side of the runner and then discharges as in the case of the Pelton wheel. The Turgo machine can allow higher speeds than the Pelton wheel and thus permits both a space saving for the turbine-generator assembly and use at lower heads, down to perhaps 100 meters. Like the Pelton, it has good part-load efficiency, but is less efficient than the reaction turbines at full load. Part-load running of the machine causes no problems.

The *Michell-Banki* turbine is another variation of the Pelton principle. It is a cross-flow machine with a radial intake, the water jets impinging on the runner fitted with curved blades at its circumference. The machine is suitable for medium heads and covers the operating range between the twin-jet Pelton wheel and high-speed reaction turbines. The turbine is relatively cheap but its efficiency is low, with about 80% at full load.

In *reaction turbines*, the runner is totally enclosed in water, which is directed on to it by fixed guide vanes, which also control the water throughput and hence the speed and power output of the machine. Action of the guide vanes, particularly full-load rejection, can produce quite large surges in the pressure conduits leading to the power plant and steps have to be taken to absorb these surges in the water conveyance system. Reaction turbines are also sensitive to the setting of the runner with respect to the tailrace so that cavitation is avoided. There is risk of cavitation also when the machines run at part load, the minimum safe loading being about 20–30% of full load. This may make it necessary to install several machines that can then be block-loaded to cope with major load changes and to maintain supplies at periods of low water availability or low demand. The principal types of reaction turbine are:

Francis turbines, which have a spiral casing distributing water radially through guide vanes to the runner from which the water is discharged axially. Larger machines generally have vertical runner axes. Smaller machines, with horizontal runner shafts, are designed for an output range of 100 kW to 10 MW and heads from about 50 meters up to 400 meters. Other designs, specifically for low-head applications, are of the "open flume" type where the spiral casing is replaced by a concrete enclosure, which forms part of the intake structure; this type is not generally suitable for small-scale use. Francis turbines are relatively expensive, but they have a low specific speed and they can operate with high efficiency—in the range of 83 to over 90%—close to their maximum rated load. They are suitable for the range of heads mostly experienced in small-scale hydropower schemes, but a disadvantage is their relatively poor part-load performance and their more stringent penstock design requirements because of the hydraulic forces and surges that their hydraulic control system can set up.

Axial flow turbines in which the water flows past a propeller forming the runner. The most widely known type is the Kaplan machine, which has adjustable blades to cope with variable heads. Other designs, such as the tubular, bulb, or rim turbines, have fixed-blade propellers, always with a horizontal shaft, whereas for the Kaplan machine the shaft is vertical. The speed control system is incorporated in the turbine runner. Kaplan machines can be used with a head of up to 50 meters, but the other axial flow machines are not suitable for heads above about 20 meters. The turbines are expensive because they require a large water throughput in view of the low heads under which they operate, but this has the advantage that the hydraulic forces acting on them are relatively small and that neither cavitation nor pressure surges in the penstocks are likely to cause difficulties. Full-load efficiencies of about 90% can be obtained and the

part-load performance of these units is generally good. The machines are unlikely to find a great deal of use for small-scale hydro power, but they can be
of value where an irrigation or water supply scheme produces only a very low
head that can be exploited for power production.

The operating ranges of the principal turbine types—Pelton, Francis and Kaplan—
are illustrated in Figure 10.3. Head, part load performance, penstock, and tailrace
configuration are generally determinant for turbine selection. Turbine speed can also
be important because the cost of the generator increases materially with slower speed.
Gearing is therefore quite common with small machines so that higher speed generators can be used. Most manufacturers produce standardized turbine types and sizes,
particularly for small-scale applications; advantage should be taken of such standardization in the interest of economy.

Control Equipment

If the hydro plant supplies an isolated network, its output must offer reasonable
frequency stability. The turbines will therefore have to be fitted with frequency gov-

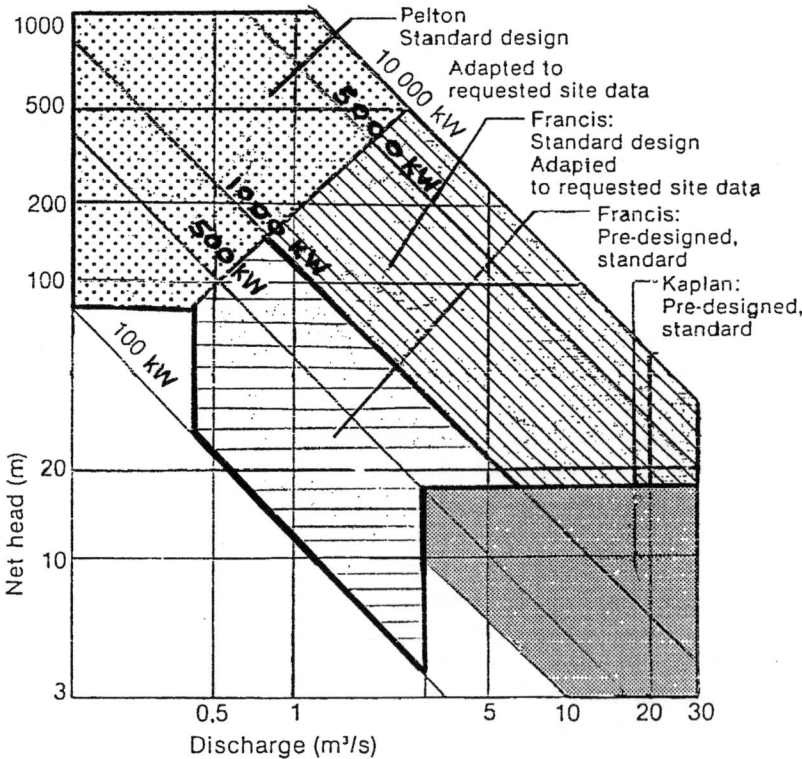

Figure 10.3 Choice of turbine.

erning of either electrohydraulic or mechanical-hydraulic type. The mechanical type is usually preferred for smaller installations, but it is greatly dependent upon competent operation and maintenance of the equipment. Governors for machines of more than one MW and of remotely controlled stations will normally be of the electrical type. To improve the frequency stability of the generating system, it is sometimes necessary to fit a flywheel to the generating plant.

The power plant design can be simplified if the plant is to be connected to a network supplied from other power units that can maintain the system frequency. Automatic frequency governing will not then be necessary and a simple hydraulic control of the turbine guide vanes for adjustment of the power output will be adequate. Methods of stabilizing the hydraulic system and of increasing the flywheel effect and in some cases even surge-preventive measures can be dispensed with. Induction generators designed for the continuous runaway speed of the turbo-generator system can be installed. A further simplification in design is possible if the power plant is to produce energy only on a block-loading basis and load regulation is carried out by other plants connected to the same network. In this case, the turbine can be fitted with fixed guide vanes and the water flow through the machine controlled by an inlet valve, which, once the unit is synchronized, is kept permanently open.

If the hydro station feeds an isolated load center or if it has to maintain supplies to an area that may become isolated in case of transmission breakdown, induction generators cannot be used and synchronous generators have to be installed instead. The generator can be made self-contained with the excitation and voltage regulation system contained in the same unit.

The generation voltage, the design of the step-up transformer, and the layout of substations and transmission line interconnections depend largely on the characteristics of the network to be supplied, its voltage and its extent. Generation voltages do not normally exceed 6.6 kV, but network voltages are governed by the amounts of power to be handled and the transmission distances involved.

TRANSMISSION

The transmission line linking the hydro power station with the load center or with the infeed point of the network supplied from the station must fulfill three conditions:

1. It must safely carry the maximum output of the station.
2. It must function satisfactorily under conditions of minimum demand.
3. Its cost must not put the economic merit of the hydro scheme in danger.

These three conditions are interdependent; both the technical characteristics and the costs of the line are decided primarily by its design. The principal feature of the line is its load transfer capacity. This depends in the first instance on the voltage and the conductor size; both are now largely standardized, although different standards are used by some of the supplier countries. The load transfer capacity of a line of given design can vary within wide limits with the operating conditions—the power factor and the acceptable voltage drop. Some typical figures for line voltages and corresponding load transfer capacities are given in Table 10.1. These figures are indicative only and are not adequate as a design basis for the line.

The load transfer capacity is expressed as a product of power flow (in MW) and length of line through which this power flow is transmitted (in km). The capacity is dependent on the cross-sectional area of the conductor (expressed in terms of the copper-equivalent area—c.e.), the voltage drop along the line, and the power factor. For equal conductor sizes and operating conditions, the load-carrying capacity of the line is proportional to the square of the line voltage. An effective way of increasing the capacity is to raise the voltage. The design of the line should therefore be targeted at a voltage level that will ensure adequate load transfer capacity at some future date, but the line can be operated initially at a lower voltage until a need for increasing its capacity arises. An economic analysis should be made in such cases:

Of the antedated expenditure incurred if the line is built for a higher voltage level
Of the capital thus lying idle during the period in which the higher voltage level
    is not needed

This analysis will help to decide whether an immediate installation for the higher voltage level can be justified.

The figures in Table 10.1 bring out the effect of the operating conditions on the load-transfer capacity for the range of line characteristics shown, especially the influence of the power factor and the benefits that can be obtained from its correction. Supplies to newly electrified areas of low-load density tend to depress the power

Table 10.1
Maximum Load Transfer Capacity of Transmission Lines

Rated Voltage (kV)	Conductor Size (sq. mm ce)*	Maximum Load Transfer Capacity (MW × km)					
		Voltage drop 6%		Voltage drop 10%		Voltage drop 14%	
		PF1.0 : PF0.85 : lag		PF1.0 : PF0.85 : lag		PF1.0 : PF0.85 : lag	
0.4	12.9	0.06	0.04	0.10	0.05	0.14	0.08
3.3	12.9	4.25	2.00	6.25	3.20	8.75	4.50
6.6	12.9	17	8	25	13	35	18
11.0	6.5	26	16	42	20	57	35
11.0	12.9	47	22	70	36	97	50
22.0	6.5	104	64	168	100	228	140
22.0	12.9	188	88	280	144	388	200
33.0	9.7	320	160	520	260	690	360
66.0	11.3	1460	670	2320	1000	3090	1545
132.0	11.3	5840	2680	9280	4000	12360	6180

*Sq. mm copper equivalent.

factor; this situation should be carefully monitored and remedial action taken to overcome inefficient power transmission. Although the load transfer capacity is greatly increased if a higher voltage drop is accepted, a high voltage drop is associated with high losses and is not desirable. A figure in excess of 10% should be tolerated only for short and emergency periods. The load-carrying capacity is limited by two further factors:

For short lines: the maximum permissible loading, usually termed the "thermal limit"

For long lines: the minimum load that can be transferred or in effect the low-load loss in the line

The thermal limit is governed by conductor size and maximum acceptable temperature rise in the conductor, taking account of the ambient temperature under which the line will have to operate. Some typical figures are shown in Table 10.2 for a temperature rise of 30° C may be excessive in some tropical situations and cause unacceptable sag in the line.

An indication of the length of line over which maximum load can be transmitted can be derived from the figures in Tables 10.1 and 10.2. At 11 kV, for example, a 12.9 sq.mm c.e. conductor will permit transmission of 8.6 MW at unity power factor and 10% voltage drop over 8 km. The transmission distance will quadruple with the same load at 22 kV, or double with transmission of twice this load (17.2 MW), which is already above the maximum output envisaged for a small-scale hydro scheme. Taking 15–17 MW as the upper limit for this scheme, voltages above 22 kV will be required only for transmission distances in excess of about 18 km. Cost considerations will, however, tend to restrict both length and capacity of the line.

The low-load loss amounts to approximately 1 kW per km under dry weather conditions but can rise to 3 kW per km in humid and polluted atmospheres. Such figures are barely significant for the distances here under consideration. Even for a

Table 10.2
Maximum Permissible Loading for Short Transmission Lines

Rated voltage (kV)	Conductor size (mm²ce)*	Maximum loading (MW)	
		PF 1.0 : PF 0.85 lag	
0.4	12.9	0.31	0.27
3.3	12.9	2.60	2.20
6.6	12.9	5.10	4.40
11.0	6.5	4.80	4.10
11.0	12.9	8.60	7.30
22.0	6.5	9.50	8.10
22.0	12.9	17.20	14.60
33.0	9.7	20.0	17.0
66.0	11.3	45.7	38.9
132.0	11.3	91.5	77.7

*mm² copper-equivalent.

50-km line operating at 22 kV, the minimum load that can be transmitted under unfavorable climatic conditions will be only about 4% of the transmission capacity.

The detailed line design and tower arrangement will depend primarily on local conditions and on the terrain. Ambient temperature and expected temperature variation, exposure to wind, and elevation above sea level are important parameters for the design. Wood or concrete towers can be built in many places and are cost-effective, especially if local materials can be used. Steel towers are not advisable below 33 kV. A single circuit does not provide a firm transmission capacity and entails a risk of loss of supply in case of line failure. An examination is necessary for deciding in what circumstances the expense of a second circuit is warranted. Transmission line costs depend on the design parameters and the nature of the terrain to be traversed.

PROJECT IMPLEMENTATION

Design

The design of a hydro power scheme is usually undertaken in three stages to a successively greater degree of refinement.

1. In the project identification and prefeasibility stage in which a conceptual arrangement of largely standardized components, based on a site reconnaissance only, is developed and costed for the purpose of a first appraisal of the merit of the scheme.
2. In the feasibility review stage in which the selection of components is refined and related to the particular conditions under which the project is to operate. A composite solution is then produced in the form of an outline design, which is costed and used to establish the technical and economic feasibility of the project. This information is presented: to government and other agencies responsible for authorizing implementation of the project and for granting the necessary licences and permits; to financial institutions and potential investors for appropriating funds for the project.
3. On securing authorization to proceed and funding, a final design is prepared in sufficient detail for the drawing-up of specifications and initiating the tendering procedure. Additional information required for firming-up the design parameters should be available at this stage and incorporated in the design to ensure that this does in fact present the best technical solution. The principal project components, as well as their composite arrangement, are optimized for best performance at least cost. Performance is measured in terms of both satisfactory functioning and long-term reliability and cost in terms of both initial investment and recurring expenditure throughout the useful life of the asset.

This three-stage approach tends to be expensive for a small scheme and steps for rationalizing the design effort are worth considering. Since the design work for both the prefeasibility and the feasibility stage is intended to provide no more than information for decision taking, the design work for these two stages may be combined and a single outline design prepared from a more careful assessment of the project parameters, which is sufficiently accurate for a feasibility review. The data derived from a more thorough investigation of the project parameters are then used for developing the final design. This approach could reduce preinvestment expenditure by as much as 30%.

The quality of the final design is important for ensuring that the scheme meets the objectives set for it and for avoiding modifications during the construction phase that may be costly to put into effect and may unduly delay completion of the project. The design should be simple, reliable and cost-effective, and fully appropriate to local conditions in the widest sense. It should cause minimum disturbance of the environment. It should ensure satisfactory operational performance of the whole scheme. It should permit maximum use made of local sources of labor and materials. Some latitude should be left to the tenderers to put forward suggestions for simplication of the civil works and standardization of the electro-mechanical plant and equipment that could result in substantial savings in construction costs. The benefits of standardization can be quite substantial for small schemes.

Specifications and Tender Documents

Specifications stipulate the performance requirements and the contractual terms for the construction effort and are essential irrespective of the parties involved in building the plant—outside contractors, manufacturers, or direct labor. The specifications, supported by the final design, must convey all the information needed for tendering purposes in a detailed and clear manner and permit the tenderer to put a realistic price to his proposals. Bills of quantities should be drawn up for the civil engineering structures detailing the materials and quantities to be used. An estimate of unit and total costs is also required for evaluation of the tenders. Detailed construction drawings can often be dispensed with, particularly where standardized designs are to be used, but careful monitoring of the contractor's work is necessary to keep costs in check, especially where the price of the contract is adjusted to the work effort entailed (the "cost plus" arrangement).

Projects in developing countries should make maximum use of indigenous resources of construction materials, manufactured equipment, and labor; the specifications should be phrased accordingly. Construction materials range from basic substances such as rocks and stones, concrete aggregate, sand and wood, to steel, bricks, tiles, and explosives. Manufactured items include wire and structural steel and mechanical and electrical equipment, depending on the degree of development local industry has reached. Simple components such as pipes and gates can often be fabricated locally provided adequate technological guidance is made available. All manufactured equipment has to be inspected before it is installed on site to verify that it complies fully with its specification and intended purpose.

The specification should set out a construction schedule that allows for proper interphasing of all activities on site and informs the tenderer of the starting and completion dates for his input to the project. The construction schedule will have to harmonize with the draw-down or dispersement schedule of project funds and the payment schedules for the contractors.

Project administration for a small scheme is simplified if the number of separate contracts is reduced to only two or three, say one contract each for civil works, electro-mechanical equipment, and transmission. A larger volume for each supply package also makes the contract more interesting for the tenderer and often leads to a keener price. Such simplification may be more difficult if local contractors participate in bidding for the work; these contractors may not have the resources to deal with a single contract containing a number of separate project elements. In such cases, the developer may find it possible to bring several contractors, local and foreign, into a joint venture or subcontracting relationship and frame the specifications so as to make this possible. The developer should ask for sufficient financial information in the tender documents to reveal any special mark-up or margins in the tender price that could arise from a joint venture or subcontracting arrangement.

General conditions of contract, including legal requirements, have now become largely standardized and it is advisable to use general forms prepared by engineering and contracting federations and widely recognized by international financing institutions.

Construction

Procedures for tender evaluation are now well established. Financing institutions usually like to be involved in the evaluation process, or at least kept informed, especially in cases where financing includes an element of aid or concessionary terms. Evaluation procedures are often laid down by the institutions. Changes in the financing arrangements may be necessary if the tender prices submitted deviate substantially from previous estimates. The tender price alone should not be determinant; the tenderer should demonstrate proper comprehension of the specifications and should be able to satisfy the developer that he has the capacity to carry out the work entrusted to him to the required standard and within the required timeframe in full compliance with the conditions of contract. He should show proper understanding for the local conditions and for the purposes and objectives of the project. He should make best use of local materials and labor and offer adequate facilities for technology transfer and training.

Equipment used for construction and for transportation of construction materials to site should preferably be of the same type as equipment used elsewhere in the country for reasons of economy and operational simplicity; such equipment will then also be available later on if required for maintenance and repair work. Use of the equipment should cause least disturbance to the local environment.

The extent of mechanization employed depends on the state of development of the country in question, the location and accessibility of the site, and the complexity of the engineering tasks to be performed. Where there is an ample supply of local labor, a high degree of mechanization may not be appropriate. Installation of the electro-mechanical equipment should be supervised by the manufacturers' erection personnel who should also witness the final start-up and acceptance trials of the plant.

A site organization has to be set up to:

- see that progress schedules are adhered to
- deal with modifications arising during the construction phase
- supervise the construction process, ensure that it complies with the specifications, inspect and test completed work, and certify it for payment. This organization must remain in being until the project is handed over to the ultimate operator

11

Operation, Management, and Training

OPERATION

Plans for operating the hydro plant should be made as soon as construction has been decided on. Local operating staff must be fully familiar with the characteristics of the plant and the purpose it is to serve. They must therefore be recruited and trained at the beginning of the construction period and must be available on site throughout the construction phase.

Power plant operation is governed by the type and size of the plant, the characteristics of the load and of the network supplied, the plant design, the degree of automation, and the organizational structure within which the plant is to function. The operating rules responding to these requirements are laid down in the first instance by the owner and manager of the plant.

If the hydro plant forms an isolated development, its operation will be governed by water availability and electricity demand. The operator will have the task to bring these two factors into balance as far as this is possible. Many schemes suffer from inadequate water runoff during the dry season and generation will then have to be restricted or abandoned altogether. The operator will have to pursue a supply policy that allows essential demands to be met for as long as possible and inessential demands to be shed in accordance with the operating rules laid down. The supply situation will be eased if the hydro scheme incorporates some storage that can help to improve the balance between supply and demand, but small schemes will not usually have enough storage capacity to permit generation during critical periods of low runoff. Optimum management of storage for maximum load coverage is an important task for the operator.

The problem of meeting the characteristics of the load at all times is greatly facilitated if the hydro station feeds into a network also supplied from thermal plant—in most cases probably a diesel plant—which can compensate for shortfalls in hydro generation. The task of the operator will, however, become more complex because he will now also be responsible for the load allocation between the hydro and thermal station and will have to ensure that the combined operating costs are a minimum.

112

This means in practice that the incremental operating cost of the thermal plant, which depends primarily on the fuel cost, has to be minimized since the incremental cost of the hydro plant is very small. Operation of the power plants and of the transmission and distribution network should preferably be in the same hands in a small system and should include inspection and simple maintenance of the facilities, preferably linked to the resources of a larger organization for major maintenance and repair.

Fully automatic operation of a small hydro station presents no technical difficulty, but the saving in personnel is small because regular supervision and inspection of the plant will be necessary in any case and this requires local staff. The personnel requirements for full operational manning of a small station amount generally to no more than one person per shift. Where a hydro and thermal station connected to the same network are located within a reasonable distance of one another, operating staff can be shared and the operation of the hydro station automated. The principal routine tasks to be carried out in the hydro plant, irrespective to automatic or manual control, are:

Inspection of the waterways, cleaning of the trash racks, and test-operating of valves and gates.
Checking of power equipment functioning and carrying out lubrication services.
Inspection of all control and electrical equipment and of the transmission lines leading to the local center.
Reading of meters and changing and filing of records.

Manual operation will in addition require supervision of the load and voltage control system, which will be automatic in most cases—and the switching of the generating unit for accepting or rejecting load. A separate organization will probably be set up to deal with consumer relations and the metering and billing cycle.

MAINTENANCE

Maintenance requirements for a small hydro plant can be divided into three components:

1. Routine or preventive maintenance, to be carried out by the operating staff.
2. Repairs—routine or emergency—and annual plant overhauls, for which the operating staff requires support from a central organization having skilled technicians at its disposal.
3. Major repair and equipment replacement, which requires the services of manufacturers or of central workshops for dealing with power plant and equipment, and of civil engineering construction personnel for dealing with the civil works.

The frequency of maintenance action depends on the size of the task. Routine inspections and maintenance need to be carried out at least weekly but daily during severe weather conditions and floods. Major inspection, minor repairs, and regular overhaul should be done annually and major repair and component replacement may be required after several of years operation, but emergency action may be needed at any time. In order to keep the hydro station in service and ensure continuity of the electricity supply—which is especially important for isolated plants—a detailed maintenance schedule should be drawn up:

> To remind plant operators of the tasks to be carried out and of the procedure for obtaining help in an emergency.
> To allocate central facilities for repair and overhaul.
> To determine which manufacturers should be approached for replacement of equipment and major components.
> To prepare an inventory of essential spares to be carried on site and at a central location.

Special attention has to be paid to damage and erosion of the waterways and of the civil engineering structures; such damage can occur quite rapidly during times of high water runoff. Particularly endangered by carryover of trash and sediment and by the scouring action of the flood water are the diversion weirs, canals, pressure conduits, and spillways, as well as the gates and valves. Damage done to these components can take a long time to repair and can result in nonavailability of the power plant for a considerable period. Another danger point is erosion of the turbine runner due to excessive carryover of sediment; it is therefore often advised that a spare runner should be purchased at the same time as the turbine so that a damaged unit can be readily exchanged.

Maintenance schedules for power plant, control equipment, instrumentation, substation equipment, and transformers should be drawn up by the manufacturers and made available with the original equipment supply. Manufacturers should also be asked to put forward suggested lists of spares that might be purchased at the same time as the power equipment and held at a central or local location, depending on the nature of the maintenance organization involved. Standardization of power plant and equipment will greatly facilitate the acquisition and management of a spares inventory; it will simplify the familiarization of the maintenance staff with the particular plant requirements and make the maintenance effort more effective.

Consumer Relations

The basic purpose of the hydro scheme is to supply electricity for purchase by the consumers. Arrangements for the purchase, metering, and billing of the electricity supplied therefore form an integral part of power plant operations, although they are usually not in the hands of the operating staff. The developer will wish to recoup the costs of the supply from consumers and will therefore wish to set tariff rates that can achieve this. The supply from a small-scale hydro plant is inherently expensive because of the relatively large capital investment involved. Consumers may find it difficult to meet the rates that will have to be charged under a cost-covering tariff, especially if they purchase electricity for the first time. It is therefore quite usual for utilities operating small local hydro schemes to average the tariff rates throughout their system and thus effectively to subsidize the consumer in a small-scale network by charges on consumers in large urban areas where he supply costs are generally lower. Tariff policy and ways in which the costs of a new hydro scheme are to be recovered should be looked into in the planning stage of the project and preliminary schedules of income and expenditure drawn up so that both the developer and financier can have some assurance that the expenditure on the scheme can be recovered.

It is importnt also to set up a strategy for metering and billing at the outset of the development. Metering equipment has to be installed, personnel trained for a regular metering and billing cycle, and an appropriate organization set up. Regular inspection has to be carried out to ensure that the metering process is effective, that the meters are properly read, and bills correctly prepared. High system losses are sometimes experienced due to illegitimate abstraction of electricity in newly developed power networks. Since most suppliers of electricity are publicly owned, the loss arising from such abstraction has to be borne by the public at large and by other electricity consumers.

INSTITUTIONAL AND ORGANIZATIONAL FRAMEWORK

The institutional arrangements for developing small hydro schemes in remote locations are often inadequate because government and utility efforts are focused on the electricity supply to the principal urban centers. Small local and rural schemes make only a marginal impact in comparison and are thus in danger of receiving only marginal attention. The management effort involved in implementing a small-scale and dispersed hydropower program, and the associated calls on the infrastructure, can be quite substantial. Management in the broadest sense requires relatively important resources of trained personnel and of finance to carry the program through—important at any rate in relation to the local economy into which the scheme is to be fitted and which, in many cases, may still be close to a subsistence level. It is necessary therefore to bring the small-scale program into focus by giving it a separate identity and developing an appropriate institutional framework for it. The institutional arrangements generally have to operate on two levels:

At governmental or central utility level, because a number of policy decisions have to be taken for even a small hydropower development, especially as regards procurement of foreign assistance and financing.

At a local level because the plant will supply only a restricted local market and will have to be operated and maintained by local personnel.

The extent of the involvement of central agencies—government or parastatal utilities—is a matter of national policy. Initiation and execution of a hydropower program requires in the first instance a central agency that can take responsibility for the overall planning of the new facility and for coordination with other enterprises for the design, financing, and construction of the new facility. The agency must also ensure that national resources are brought into play as far as this is possible, especially for the supply of equipment, construction materials, and labor. Furthermore, the agency must ensure that the plant is properly operated, maintained, and utilized.

Among the policy decisions that have thus to be resolved by the central agency are the following:

1. The question of overall rural energy development and of the role of hydropower in it.
2. The way in which the hydropower program is to be managed and financed.
3. What sources are to be approached for the design and construction of the plant and the procurement of the equipment to be installed in it, with special reference to local participation in the construction effort.
4. How the new plant is to be operated, maintained, and utilized, and what tariff policy is to be pursued.
5. How the personnel to be engaged on the management and operation of the scheme is to be recruited and trained.

The institutional arrangements must be so framed that these problems can be effectively dealt with. The success of hydropower development is greatly dependent on strong institutional support and adequate attention must therefore be given to measures for securing this support. This matter is also of importance for external aid agencies, which can only function satisfactorily if an effective local counterpart organization has been set up and is functioning.

The only marginal impact of a small-scale program can be overcome to some extent by giving it a separate identity and entrusting it to an organization specifically charged with its development. At the central level, this organization can be a department or group within the government or utility structure responsible for dealing with the issues raised above and for providing guidance to the local organization dealing with the project. There are a number of possibilities for the institutional arrangement at the local level:

A local unit can be set up by government or by the national electricity supply utility for running the plant.

A municipal or rural electric power cooperative can be set up, with guidance from the central level.

A communal or sectoral enterprise can be established by special interest groups, for example, by local industry or commerce, or by local government agencies or missions.

Private enterprise may also be involved in some cases, especially where the output from the scheme is intended primarily to supply a private sector activity.

The local organization will need to maintain some links with central government because it is unlikely that issues involving foreign currency commitments can be dealt with without governmental intervention and approval. Questions of licensing and allocation of water rights will also impinge on government authority and will have to be resolved in cooperation with central and local government. Matters of technical policy, construction arrangements, plant procurement, personnel deployment, and training can, however, be satisfactorily dealt with at a local level provided the local organization is adequately equipped to do so or can obtain external support for which government approval will probably be necessary.

TRAINING REQUIREMENTS AND PROGRAMS

The success of a hydropower program depends greatly on the availability of adequately trained and experienced personnel who have the right skills and aptitudes to play a proper role in all phases of the development and ultimately the operation and maintenance of the hydro facilities. Developing countries often have to rely on external assistance for initiating a hydro power project, for carrying it through the planning cycle, for overseeing its construction, and for ensuring satisfactory operation. Countries need to realize, however, that such assistance is not available on a permanent basis and that—sooner or later—national personnel must become involved in the project; national personnel must take it over and run it satisfactorily in the knowledge that a valuable asset has been created that will have to be properly used and maintained. If there is prospect of an expanding hydropower program, local personnel must become acquainted with the identification of resources, their progressive development, and the practical realization of project proposals. Reliance on external assistance should be gradually reduced to key issues of technology transfer so that a greater degree of self-sufficiency can be developed, a national technology base built up, and uniformity of approach and design appropriate to national requirements achieved. Training of personnel, with this objective in view, must be taken into account at the very beginning of a hydropower program, even if only a single plant is under consideration initially, and it must be accorded due priority in the development plans.

Bearing in mind the size and decentralized nature of small-scale hydropower, personnel formation should be envisaged at two levels:

1. Training of key personnel at the professional level who will occupy a central position in government or in the national electricity supply utility with responsibility for guiding and supervising the planning and development of small-scale hydropower nationally, advice government, carry out government policy, and ensure coordination with external agencies, consultants, and contractors brought in to implement a specific development plan. Such personnel will probably receive suitable training abroad, individually or in groups, with utilities,

consultants, and manufacturers, through fellowships and study tours supplemented by attendance at courses, seminars, and workshops. Useful training can also be obtained through secondment of personnel to developing utilities in the same region, which may face similar problems, or indeed through an exchange of personnel with such utilities.

2. Training of local personnel at the technician level who will be employed on site and in the plant, initially for assisting during the construction phase and ultimately for operating and maintaining the plant. These site personnel will receive primarily on-the-job training and will familiarize themselves with the technical and operational characteristics of the scheme and of the equipment through their presence on site throughout the construction period by arrangement with the contractors concerned. It may also be useful on occasion for site personnel to receive on-the-job training in neighboring utilities either by secondment or through staff exchange, especially at the beginning of construction when site staff should be trained for active participation in the more complex plant installation activities.

Once the training requirements have been defined and suitable personnel identified, it is necessary to formulate appropriate training programs and to arrange financial and technical support for them. The basic elements of the training program are:

1. Determination of training needs for both individuals and groups. This entails an assessment of the performance required of the trainees when their training course has been accomplished.
2. Establishment of training aims and objectives, i.e., of targets that the training course is to achieve. This involves an appraisal of the kind of training that is to be undertaken and the selection of training methods, materials, and locations.
3. Implementation of the training program.
4. Review of the results of the training course, appraisal of the extent to which the aims and objectives originally set have been achieved, and adjustment of the training strategy where necessary.

Although the scope and flexibility available to developing countries is necessarily limited, careful personnel planning should nevertheless be undertaken so that the requirements for trained personnel can be met in the best possible way. Training is one of the principal fields for which technical assistance is available from many sources. The developer is therefore urged to secure such assitance as soon as a particular project is initiated and as an integral component of other assistance received toward this project. The shape and scope of the training activity should be relevant to both national and project needs. The developer should keep in mind that the training activity will have to aim at long-term objectives and that it must not be framed too narrowly toward immediate needs. For proper effectiveness of the training effort, the developer should ensure that:

Personnel are made available for training at the appropriate time; training must not be impeded through delays in secondment of personnel.

Personnel selected for training are capable of absorbing such training and of carrying out the tasks for which the training has been given.

Personnel remain assigned to the tasks for they were trained; the period of assignment should be for a reasonably long time so that the training effort is not squandered through frequent changes of duties.

The advice and assistance of agencies sponsoring and supporting training programs should be freely sought when establishing such programs. The general nature of the type of training involved in a small-scale hydropower scenario is much the same throughout the developing world and aid agencies will therefore have accumulated considerable experience in the formulation of programs that can meet both general and specific requirements. An effective way of securing training support is also through "twining arrangements" in which the power utility of the developing country forms a close link with a similar utility in a more advanced country, which can offer extensive training facilities. "Twining" between developing countries can also be useful in some cases.

Where more extensive hydropower facilities are envisaged, training can be used to form a pool of experienced personnel who can move from project to project and help to set up local training centers nationally or on site. Technological self-sufficiency can thus be developed with only casual support from abroad. This approach will be particularly suitable for the formation of technician and artisan grades of personnel.

INTERNATIONAL COOPERATION

The range of experience and skill available for the initiation and development of hydropower program can be greatly strengthened and expanded through cooperation with other parties who are active in the hydropower sector. The assistance derived from such cooperation can be particularly important in the early stages of program. Local experience may then not yet have sufficiently developed, but crucial decisions on policy, technology, economics, and finance may nevertheless have to be taken. Local personnel may also have to be prepared for the task of operating and managing new and unfamiliar facilities. National cooperation may help to some extent to overcome such problems in places where hydropower has already been established in a different area and where an exchange of experience with a newly developing area may then be possible but, in most cases, contact outside the country will have to be looked for.

Access to international experience can be of considerable help in all stages of hydropower development, from the initial planning of a single scheme or a more extensive program to the final operation and maintenance of the power plants and associated transmission lines, their management, and commercial aspects including tarification and governmental, interutility, and consumer relations. International cooperation is available from a wide range of bodies, but primarily from the following.

1. *Multilateral organizations* operate worldwide or regionally on a governmental or semigovernmental level, but basically dealing with public sector issues. Principal among these are the United Nations and its central and regional agencies as well as the World Bank and the regional development banks. Multilateral organizations formed by particular interest groups can also offer valuable cooperation; among them are bodies such as the OPEC Fund for International Development, the Latin American Energy Organization (OLADE), the Organization for Economic Cooperation and Development (OECD), the European Economic Community (EEC), and others. The important element here is that support and assistance are provided by a group of countries that combined to establish a central body carrying out their particular aid programs. Assistance provided by multilateral agencies is generally untied, which means that personnel carrying out the assistance program or any goods and services associated with it need not necessarily come from any particular country; the recipient of assistance can often participate in selecting suitable sources, although, where such assistance is given from a more restricted group of nations, the choice of assistance personnel is often limited to the nations participating in the particular assistance activity.

2. *Bilateral organizations* provide support on a country-to-country basis. The most widely developed form of this kind of cooperation stems from industrialized countries that are prepared to make their experience available to the developing world, generally way of aid. Cooperation with a bilateral source is almost invariably tied to services from the country in question, but where the donor has the right kind of experience, such services are usually of very high quality. The recipients will, of course, appreciate that the motivation for the technology transfer available to them by way of bilateral cooperation is not always charitable or humanitarian but quite often political or commercial.

3. *Technical cooperation between developing countries* (TCDC) offers an effective way of sharing experience between nations facing similar problems. Originating from cooperative ventures between individual countries, the concept has spread to a group of countries cooperating through exchanges of personnel and regular meetings, workshops, and seminars at which matters of common interest are discussed and developed. TCDC is a valuable extension of the aid made available by multilaterial by bilateral agencies from industrialized countries.

4. *Nongovernmental organizations* (NGOs) have not been set up on the basis of agreements between governments, but many are active in the field of international cooperation. Such organizations are mainly formed by professional

and technical interest groups that sponsor seminars, workshops, and conferences for the discussion of many aspects of the hydropower sector and can provide advice on an ad hoc or regular basis to developing countries. The services of experts can also be made available through nongovernmental organizations; such services can play a useful role in providing expertise in specialist fields. Financial restrictions often limit the scope available to such organizations unless government support is provided for them.

The objectives that can be served through international cooperation cover principally:

- technology transfer through all forms of technical assistance
- exchange of experience through secondment of personnel, group discussions, seminars, workshops, and conferences
- all forms of training of technical, administrative, and managerial personnel both abroad and on-the-job
- joint activities and operations for the purpose of personnel formation and the sharing of experience

Part III

Small Energy Packages

12

Introduction

NEEDS

Over half the world's population live in rural areas of developing countries, which are for the most part poorly served with central electric power and other forms of conventional mechanical and electrical energy. Currently their needs are met, to the extent that they are met at all, by small internal combustion engines and engine generators (diesel or gasoline).

Small internal combustion engines, although relatively inexpensive in view of their mass production for automotive purposes, have a limited life and require skilled maintenance, which is not always available in remote areas. Furthermore, fuel costs rose sharply in the 1970s and early 1980s and even though they have recently declined on the world market, they remain a considerable burden to oil-importing developing countries so that prices of petroleum products remain high in most parts of the world.

These small energy sources typically cover a range of from 1 kW up to 1 MW (1,000 kW) or more, and for certain purposes (such as domestic lighting) there are requirements for sources downward to as little as 10 W. Many of these small energy sources are candidates for eventual replacement by the new and renewable sources of energy that have been under development in recent years, especially since the 1981 United Nations Conference on New and Renewable Sources of Energy in Nairobi.

RENEWABLE ENERGY

Nevertheless, development and diffusion of small renewable energy sources based on solar energy, wind energy, hydropower, or biofuels have been slower than many expected. The reasons are varied and complex and often different for different technologies within a country, or for the same technology from one country or region to another, and sometimes even from one time period to another.

In this respect a case study of small wind pumpers would be rewarding. Some six million are said to have been installed in the United States since their introduction in 1860 and many thousands of these are still in use. This is also the case in some European countries and in Argentina, Australia, southern Africa, and especially in China where the numbers are said to approach 500,000. In many developing countries, large numbers were introduced in colonial times, but, following independence, have fallen into disrepair. In a few of these countries, steps are being taken to repair and rehabilitate existing wind pumpers and to promote the local manufacture of improved ones using modern materials. These efforts have, in most cases, had only a limited success, for various reasons, including:

Insufficient resource (wind) where it is needed.

Lack of familiarity with the technology.

High investment ("front end") cost.

Absence of credit facilities available to the potential users.

Foreign exchange restrictions (where there is no local manufacture).

Absence of services for repair and maintenance.

Competition from small internal combustion engine pumps, which may be more affordable in the short run (because of low investment cost) even though they have higher "life cycle" costs.

Competition from solar photovoltaic (PV) pumps, because of their simplicity and minimal maintenance requirements, even where they may be more costly than windpumps (this is becoming a factor in Australia, for example).

APPLICATIONS

The above case study on wind pumpers leads to the question of applications. One of the most important applications for small energy sources is pumping water for habitations, animal watering, crop irrigation, and land drainage. This application is particularly well adapted to stochastic (randomly varying) sources such as solar and wind energy since, within certain limits, the resultant varying product throughput is acceptable, although even in this case there may be some extra expense for a storage rank or reservoir to reduce the fluctuations.

Undoubtedly the most important application for small energy sources in rural areas is as a multiuse facility, or minigrid, to supply electric power for lighting of public buildings, streets, shops, workshops, and homes and for powering small tools and appliances such as fans, radios, television.

In many cases the engine generator will continue to be the source of choice. However, in special circumstances the existence of a suitable nearby stream could make a microhydro unit preferable. In other cases the existence of an agro-industry with combustible wastes or of a forest resource that can be exploited on a sustainable basis might make an engine generator with biofuel (gasifier plus dual-fuel internal combustion engine; or steam engine or small steam turbine; or air—stirling cycle—engine) an attractive choice.

In windy areas (typically islands and coastal areas) a hybrid wind/diesel generator system (with wind supplying up to as much as 90% of the load, according to design) may be considered. Currently control problems with small wind/diesel systems remain not fully resolved, but the situation is likely to improve in the near future, making this an increasingly viable option.

Finally, solar photovoltaic (PV) systems may be considered; however, these will need battery storage, especially if the main demand is for lighting and evening use. Present prices of solar PV and of storage batteries make them a viable choice only for the smallest loads of a few kWh per day where they are competing with very small diesel generators (2–4 kW).

On a still smaller scale there is the wind or solar PV battery charger with 1 kW or less nominal power and providing electrical energy usually for a single home or building. Here the unique advantages of solar PV—namely, simplicity and reliability as well as modularity in the sense of being available in an almost infinite range of sizes from the smallest to very large—makes itself felt. In fact, in this case solar PV battery chargers are, to a considerable extent, competing with kerosene lamps rather than with engine generators. The main factors holding back their widespread dissemination is probably their high first cost combined with lack of credit facilities in the rural areas, and lack of awareness of their possibilities.

Finally, small energy sources are required for a variety of specific purposes including ice making for fishing villages, water desalination, vaccine refrigerators for rural clinics, telecommunication (including roadside telephones, telephone repeaters, and HF communication links), aviation and maritime navigation aids, railway signalling, cathodic protection of pipelines, and so on.

CHOICES

In all of the above examples the energy source of choice, if available at a competitive cost, would undoubtedly be electric power from a central power supply. However, even when a high voltage transmission line passes quite close to a small community, the cost of transformer substation and of local distribution lines can be quite prohibitive unless there is a substantial local demand. However, there has been a recent development of power voltage transformers capable of directly transforming high voltage (to 245 kV) down to domestic voltages (230 V single phase or 400 V three phase) with outputs of 25 kVA or 50 kVA per phase, which may ease this situation.

If power supply from a central grid is out of the question, it is then necessary to evaluate the cost of an engine generator, which is usually the next choice for power demands of 1 kW and up. Account should be taken of possible alternatives such as microhydro, if there is a suitable stream, or the use of producer gas (from biowastes) as a fuel in a spark ignition or duel-fuel (diesel) engine. For particular applications where the load may be flexibly adapted to a fluctuating power supply, as in water pumping, ice making, water desalination, and vegetable oil extraction, consideration should also be given to wind or solar energy. For the smallest power requirements below 1 kW, solar or wind battery chargers may be the preferred energy source. To assist in evaluating these alternatives, see the life cycle costing procedures at the end of Part III.

13

Energy in Rural Areas Third World

This chapter reviews the basic data on the rural areas in Third World countries to provide a general framework for what follows. The problems of supplying energy in rural areas, and the problems of electrification in particular, are then tackled. A rapid analysis of these problems identifies the limits of conventional rural electrification and the need to examine other alternatives.

RURAL POPULATION GROWTH

In round figures, about 2.5 out of 5.0 billion people are living in rural areas in what is commonly called the Third World; in other words, about half of humanity consists today of the rural inhabitants of the developing countries.

Despite the prodigious urban growth observed throughout the Third World in the past three to four decades and despite the massive migrations driving the rural inhabitants to the cities, more than two-thirds of the Third World population are still villagers. And the rural population of the Third World is continuing to grow inexorably, albeit slowly.

Only in the first decade of the twenty-first century will the urban population probably exceed the number of rural inhabitants in the developing countries. And it will undoubtedly be fairly far into the twenty-first century (although any forecast concerning such a distant horizon is certainly unreliable) before the rural population in this part of the world will begin to decline. This situation prevailed in many of the now industrial countries before the end of the nineteenth century, and only occurred in the other industrial countries in the learly twentieth century.

These overall figures, however, conceal a widely diversified reality. Barely one-third of the population is rural in Latin America, whereas about three-quarters of the population of Asia and Africa south of the Sahara are rural. With a 55% rural population, North Africa occupies an intermediate position.

ENERGY IN RURAL AREAS

The large majority of these 2.5 billion rural inhabitants essentially consume traditional energies, not commercialized but collected by each household: wood, firewood, plant, and animal wastes. These fuels serve to cook meals, possibly to heat the water, and dwellings as required by the rigor of the climate, and also to satisfy the needs of rural craftspeople: smithies, pottery kilns, brickworks, traditional beer producers, and others.

We are familiar with the problems raised by the supply of traditional energies and the increasing difficulties faced by the rural populations in ever-increasing areas of the Third World in finding the daily fuel necessary to cook their meals, and the pressure that this collection exerts on the environment. The great majority of villagers have no energy other than that of their own arms and of a few domestic animals to work the fields, to process crop produce, and to draw water.

The large majority of the rural population does not have other than traditional means for their mediocre lighting. An oil lamp produces about 10 lumens. A good-quality, kerosene-burning hurricane lamp yields 60, whereas a modest 40-watt electric bulb or a more efficient 10-watt fluorescent lamp generates approximately 500.

In Africa south of the Sahara, the traditional energies supply nearly all the energy consumed by the rural population. In the Indian subcontinent (India, Bangladesh, Pakistan), they supply about 90% and in these areas, animal wastes alone provide half of all energy consumption. In a more advanced country like Mexico, according to 1975 data, the traditional energies still supply 80% of the consumption of rural households.

These orders of magnitude highlight the very limited place occupied by modern commercial sources of energy in the developing rural world: petroleum products (butane and kerosene for cooking, kerosene for lighting) and electricity. In actual fact, by virtue of its many applications, electricity is seen as an essential element of a certain quality of life, and also a major factor for development. Hence it is worthwhile examining further the place of electricity in rural areas in developing countries.

THE SITUATION OF RURAL ELECTRIFICATION

The first question that comes to mind concerns the number of Third World villagers who really have access to electricity. It is extremely difficult to answer this question, or even to provide a rough approximation. The available data are incomplete and often unreliable. Not all countries have the same definition of what is called "rural electrification." Some consider that apart from the capital and a number of large towns, *all* other electrification is rural; whereas it actually only concerns built-up areas whose urban character is undeniable in the universally acknowledged sense (areas of more than 5,000 inhabitants) and practically does not exist in the rural areas in the true sense of the term.

Other countries publish figures for the population in the rural areas covered by electricity distribution networks. These are often estimates of the population that *could* have access to electricity *if* it possessed the means. In actual fact, a small part are effectively customers of the utility, whereas the remainder are far from possessing the means to pay for a connection or even modest consumption. With these reservations, to what extent is a significant statement possible?

In sub-Saharan Africa, a figure of 4% of the rural population with effective access to electricity is generally acknowledged and seems plausible. For many African countries, the rate of rural electrification is close to zero. This applies in particular to the least advanced countries, but the rural population of many countries that have already initiated development is barely better provided. Only a few countries like Mauritius and the Ivory Coast appear to have a rate of rural electrification above 10%.

In North Africa, a rate slightly above 20% probably reflects reality fairly closely, and this average covers rather different situations. In Algeria, about one-third of the rural population enjoys electricity, but the proportion is lower in Egypt at 20%, and even lower in Tunisia and Morocco (10% and 6%, respectively).

In Asia (not including China), the home of slightly under 1 billion rural inhabitants out of the 2.5 billion counted in the Third World, no more than 180 million enjoy electricity, with very sharp disparities. The large majority of South Koreans are connected to the grid, whereas only 1% of the inhabitants of Sri Lanka have access. The rate of rural electrification in these countries in general ranges between 30% and 45%.

The literature informs us of a 69% rural electrification rate in India and 50% in China. In India, this is clearly an illustration of the proportion of rural households that could benefit from electricity if they possessed the means. In actual fact, the level is very probably much lower. An Indian survey shows that in villages electrified for several years, the rate of connection of the households is about 20%. This rate is probably a maximum limit for the proportion of the population effectively enjoying electricity. The same probably applies to China, but it is impossible to determine clearly the proportion of the rural population that really has access to electricity. In Latin America, the claimed rate of rural electrification is 27%, and some countries have more than 40% of the rural population "electrified" (Uruguay, Mexico), whereas others are still at a very low level: 5% for Brazil, 10% for Bolivia. On the whole, a plausible order of magnitude for the rural population of the Third World is about one out of five enjoying access to electricity.

If we now attempt to evaluate the electricity consumption in the rural areas of the Third World, the problems faces are just as difficult. In the early 1980s all the developing countries combined consumed only 13% the total electrical energy consumed throughout the world. Yet, most of this energy was and continues to be consumed in the urban zones. In India, where rural electrification programs have been implemented for several decades, per capita consumption in the early 1980s was 15 times higher in the urban than in the rural areas. Assuming that this proportion applies in the other regions of the Third World, this means that about 1% of the electrical energy generated throughout the world is destined for 2.5 billion rural inhabitants of the developing countrie. This figure of 1% appears to be a plausible order of magnitude.

RURAL ELECTRIFICATION AND ECONOMIC DEVELOPMENT

Access to electricity affords the rural population an obviously better quality of life. But what does it offer from the development standpoing? In fact, the available data suggest that no close correlation exists between the level of economic development (per capita GNP) and the degree of rural electrification.

The most striking example is undoubtedly that of the comparison of Brazil with South Korea, two "newly industrialized countries" both at the level of $1,900–2,000 per capita GNP (in 1983). They are from the standpoint of rural electrification at the two extremes of the spectrum, one with 5% of the rural population electrified, Brazil and the other with 90%. Two neighboring countries may also find themselves in quite different situations; rural electrification is far more developed in India than in Sri Lanka, whereas Sri Lanka enjoys a higher per capita GNP.

Three factors are clearly involved and explain the lesser or greater development of electricity in the countryside:

   The rural population density. It is evidently proportionately less expensive to electrify 15 million rural South Koreans spread over less than 100,000 km^2 than to conduct the same operation for 40 million Brazilians dispersed over some 8 million km^2.

   The income of the rural population and hence the income distribution between urban and rural inhabitants.

   Political determination. Rural electrification has clearly received high priority in India. In certain Indian states, priority has been to electrify water pumping in order to develop irrigated cultivation rather than to satisfy domestic needs. This is the case in Punjab, where all the villages now have electricity. In other Indian states the objective was social, as in Kerala, a relatively advanced state at the southern tip of India where all the villages have electricity.

   In Sri Lanka, a neighbor of Kerala, the concern for developing irrigation does not exist, and rural electrification has not been assigned priority. The state has been particularly concerned about the "quality of life;" for the income level in that country, the rates of school attendance and life expectancy are extraordinarily high, but access to electricity has not been considered as part and parcel of social policy.

Many developing countries that have reached the developmental level that some industrialized nations held some years ago show a more advanced rate of rural electrification. It is worth remembering that rural electrification had barely begun in Europe and North America before World War I. It took place essentially in the interwar period. In 1920, in French rural departments, about 5% of the population enjoyed electricity. By 1930 this proportion had not yet reached 40%. And the final "holdouts," permanently inhabited, were only connected to grid in the early 1960s.

Although any comparison of development levels in time and space is problematic, the mean per capita GNP is the most developed countries in 1920 was higher than what it was in 1985 on the average in the developng countries. Some experts estimate that the per capita GNP was $625 (1960 value) in the industrial countries in 1920, and $835 in 1930, whereas it was $400 on the average in the Third World in 1985. The developing countries benefit from the reduction in costs due to technological progress since the time when rural electrification was carried out in the industrialized countries.

The impact of rural electrification on development has often been disappointing. A survey carried out by the ILO concludes that "based on the examination of the literature on the subject and statistics, the advantages of rural electrification, including benefits on the social level, tend to be overestimated while the costs are underestimated."

Most of the assessments of rural electrification projects show that these investments have not had the anticipated impact, neither on the transformation of farm production systems nor on industrialization in the rural areas. From the social standpoint, they have barely affected the most disadvantaged classes of the population. And on the contrary, one may well ask whether they have not helped to widen the gap in income distribution.

In India, where the development of pumping irrigation has been one of the major motives of rural electrification programs, the service provided has been of such mediocre quality that the number of diesel motor-driven pumps has increased spectacularly. In China, on the other hand, rural electrification appears to have had a considerable impact on the mechanization of agriculture, by facilitating the creation of many workshops for the manufacture and repair of agricultural equipment. Elsewhere, it has not played the role of catalyst of rural development that was anticipated. If it has contributed to this development, it has been only one factor among others and undoubtedly not the most decisive.

LIMITS TO CONVENTIONAL RURAL ELECTRIFICATION

In the industrial countries of Europe and North America, rural electrification started with a preliminary phase, which included:

The construction of small power stations feeding mini-distribution grids supplying a group of villages, a village, or even a hamlet or a single farm.
Simultaneously, or very shortly thereafter, the establishment of larger power stations (yet small in comparison with those of today) supplying medium voltage grids, which were gradually extended to include the large majority of the rural communes.

In a second phase, the small and medium-size power stations disappeared and were superseded by very large power stations generating electricity at much lower kWh cost and supplying the rural distribution systems through a high voltage grid interconnected on a nationwide, indeed continental, scale.

The process was substantially repeated in all the industrialized countries, with a few significant differences. In Europe, where the rural population densities are relatively high, the techniques adopted for electricity distribution in the rural areas were relatively costly in terms of capital (three-phase systems), whereas in North America, where the rural densities are lower, the techniques applied involved lower capital costs (single-phase systems).

The developing countries are committed in turn to the same process, extending the distribution networks around a central generating unit, of variable size, or around interconnected high voltage grids.

Whereas this method of rural electrification development benefits from the experience gained in the industrial countries and from advances in electricity generation and distribution techniques, it also faces major obstacles.

Capital cost in transmission and distribution systems is high, higher in the rural than in the urban areas. The situations are quite different from one country to another, but for many developing countries, this high cost is a serious restraint to electrification. This is especially true of many of these countries that do not themselves produce part of the hardware required and are forced to import it from the industrial countries. The lack of foreign exchange, the indebtedness already deemed insupportable by a large part of the Third World, is obviously a highly unfavorable factor for the implementation of new major rural electrification projects in these countries.

Nor does the observably weak impact of these projects on rural development encourage aid sources to allocate large funds to this type of project. In the African countries in particular, it is likely that a number of aid agencies will direct their future activities more toward directly productive projects than toward investments of a social character.

The delivered kWh cost to the consumer is high, due to the high capital cost for its distribution, the high kWh cost in generation based on petroleum products, and the low consumption of rural customers. In the first phase at least, the uses of electricity will often be limited to a few lighting points.

The same observation applies to the generation of electricity by small diesel power stations supplying a small number of users, and possibly a single one, owing to the constraints specific to this type of power station (maintenance costs) and the high cost of fuels and lubricants.

The national income distribution in many developing countries is such that the villagers have very low incomes, especially monetary incomes, and the kWh cost makes electricity inaccessible to a large proportion of the rural population.

The importance of these obstacles to the development of rural electrification must not be underestimated. In countries where the dynamics of evolution of the rural world is already well on its way, resources will probably be set aside to ensure continued electrification by conventional processes. On the other hand, the countries where this evolution has barely been initiated, and where only a few percent of the rural population have access to electricity, it is doubtful that the coming decade will witness any spectacular acceleration of electrification. On the contrary, all these factors militate in favor of a slow growth process, and even stagnation in the number of households connected to the grid, and of electrical power consumption by these households.

OTHER WAYS WORTH EXPLORING

Are the developing countries condemned to follow the same path as the one taken by the countries that are industrialised today? Certainly not. Whereas conventional electrical energy generation and distribution methods have registered considerable progress in the past several decades, other techniques have marked even greater advances. The use of wind energy and small waterfalls has been reduced to virtually nothing by the extension of electrical grids in the industrialized countries (although the generation of electricity by the use of wind energy and photovoltaic energy has recently shown remarkable progress in California). Yet new techniques may again make these forms of energy interesting for the developing countries, at least in certain cases. Similarly, biomass can henceforth be employed by more efficient methods than traditional combustion. And everyone knows that it is now possible to convert solar energy directly into usable electric energy.

The delivered cost of electric energy in the rural areas to a more or less extended grid (which may ultimately only concern a single user) is obviously the reference against which other alternatives of decentralized energy production must be compared. But the viability of these alternatives cannot be assessed by this comparison alone. Modern techniques for the utilization of renewable energies make it possible to:

- supply a mini or micro-distribution grid by a power station
- satisfy a localized energy need: lighting, refrigeration, water pumping, running a telecommunication unit, etc.

In each case, the problem is to find a sound match between:

- a renewable energy resource (wind, hydro, solar, biomass) that is available
- an energy need, clearly identified and solvent
- a technique that serves to satisfy one from the other, in acceptable technical, economic and even sociological terms

Hence the problem is much more far-ranging than a simple comparison of production cost of kWh generated and available to the consumer.

14

Rural Electrification by Extension of Interconnected Grids

CHARACTERISTICS OF RURAL ELECTRIFICATION

The term "rural electrification" concerns the supply of electricity to low and very low density areas. It is traditionally achieved in two ways: by the installation of generators independent of the grid (diesel sets) directly at the consumption site (village, farm, small industry, dispersed dwellings), or by the extension of the interconnected electrical grid (according to the World Bank, this latter technique accounts for 80% of rural electricity distribution). For electricity distribution, rural areas are distinguished from urbanized areas by some fundamental aspects:

Sites to be electrified are often several kilometers from the existing medium voltage (MV) network.
There is much lower population density.
Electricity consumption is much lower than the average urban consumption.

HIGH CAPITAL COSTS

The characteristics of rural electrification described above result in an increase in capital costs of rural projects (per new customer) in comparison with urban projects because:

The great distance of sites to be electrified entails the installation of MV lines from the grid over sometimes significant distances (2 to 3 km on the average in India; 12 km in Brazil). As a rule of thumb, stability problems limit extensions of grid to a distance in km of not more than double the line voltage in kV.
The low population density in comparison with urban sites requires the installation of longer low voltage (LV) lines per new customer.

The type of terrain encountered may sharply increase network extension costs, particularly in relatively inaccessible areas (mountainous or thickly wooded regions).

Project cost is also highly sensitive to the ability or inability of the country concerned to produce the electrical hardware required (conductors, poles, transformers) instead of importing it, but this factor exerts the same influence on rural and urban projects. In India, which produces a large part of its equipment, the cost per kilometer of LV line is $2,100, whereas in Benin, which imports it, the cost is $16,000.

The constraint of the terrain, the scope of industrial capability (imports or domestic product), the technical qualifications of the utilities, and the scale and substance of the programs (India, for example, is developing a program strongly directed toward irrigation, thus requiring fewer individual connections) explain the very wide disparity between the costs of the projects, as indicated by Table 14.1 The disparity can be explained partly by the difference in size between the villages and hence by the number of potential users concerned by the project, but this is not the only cause.

FACTORS INFLUENCING CAPITAL COST

The population density or the number of potential customers per km of line is the primary factor having a strong influence on electrification cost. Table 14.2 compiled from Paraguayan projects, clearly shows a close correlation between the connection cost of a new customer and the user densities. This characteristic is obviously not specific to developing countries, as these correlations also exist in the industrialized countries. For densities greater than 20 potential users per km of line, the cost per user is around $600, but for lower densities with about three customers per km of MV and LV line, the mean cost is much higher than $3,000 per user. In actual fact, as rural electrification extends into the hinterland toward more sparsely populated

Table 14.1
Costs of Rural Electrification (in 1985 US$)

Countries	Cost per village*
India	20,000 to 30,000
Bangladesh	70,000
Zambia	73,000
Tunisia	83,000
Bolivia	105,000
Pakistan	80,000 to 110,000
Indonesia	125,000
Benin	300,000
Burkina Faso	320,000

*Costs are orders of magnitude derived from the national local averages for villages of from 1,000 to 5,000 inhabitants. In the case of Benin and Burkino Faso, they represent secondary urban centers with 10,000 to 15,000 inhabitants.

Table 14.2
Sensitivity to Density in Paraguayan Projects (costs in 1982 US$)

Project	No. of potential users	Lines (km)		Density* users/km/ LV + MV	Cost potential user
		MV	LV		
Concept. 1	5,208	100	60	33	500
Concept. 2	795	15	21	22	590
Itapua 4	794	18	23	19	650
Itapua 2	337	16	12	12	1,000
Caazapa 1	208	51	16	3	3,650
Alt. Para 1	205	46	14	3	3,300
Alt. Para 2	378	86	25	3	3,380
Neemburu 2	166	50	10	3	4,210
Neemburu 4	133	44	8	3	4,200

*For densities greater than 20 potential users per km of line, the cost per user is around $600, but for lower densities with about 3 customers per km of MV and LV line, the mean cost is much higher than $3,000 per user.

zones farther from the grid, the number of customers per km of line decreases, so that the costs of electrification of each new customer increase.

The Algerian national electrification plan indicated that between the conception of rural electrification in 1975 and its planned completion in 1987, the average connection cost of a new customer was expected to rise from US$750 to $2,500 owing to the drop in density of the new lines (Figure 14.1). On the average, between 1975 and 1978, 1 km of LV and MV lines served to connect 26 new customers. In 1983 the mean density of new connections fell to 17, and the national plan called for this figure to drop to 10 in 1987. These sharply increasing costs do not concern only a few remote rural areas. According to the electrification plan (1979), 50% of the centers remaining to be electrified (125,000 households) will have connection density/km of line less than 10.

Thus, after the connection of the rural customers grouped in the large villages (over 1,500 inhabitants) and located in readily accessible areas—population of variable size according to country—rural electrification enters a phase of sharply rising costs.

GRID OPERATION

Operating costs also increase with the "rurality" or "low density" of the areas to be served, for various reasons:

As the customers are located farther from the distribution centers, it becomes more difficult to check pilferage, and routine maintenance and meter readings become more costly.

Service interruptions are more frequent (and longer due to distance) because the line length exposed to the weather, vegetation growth, etc. is greater.

Finally, and above all, line losses become very high (increasing with the square of line length).

Although rural electrification is still little developed in the Third World countries (see page 00), is is clear that their distribution problems are already acute: 50% of these countries "lose" more than 15% of the generated electricity (see Table 14.3). The spread of rural electrification, by augmenting the above difficulties, will undoubtedly aggravate this situation.

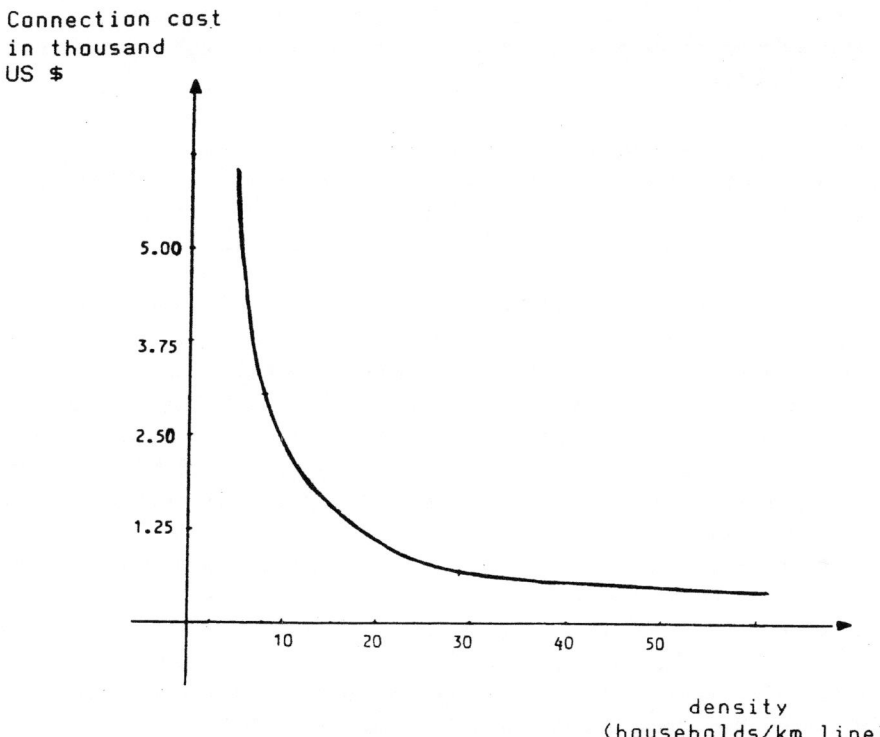

Figure 14.1 Connection cost as related to density—Algeria. (*Source:* Algerian National Electrification Plan)

Table 14.3

Electrical Losses in Transmission and Distribution Grids of the Developing Countries

Losses (% of generation)	% of countries concerned
0–10	17
11–15	33
16–20	21
21–30	21
31–40	8
	100

Source: World Bank 1983.

TYPE OF CONSUMPTION AND kWh PRICE

Two types of consumers account for most of the rural demand, domestic and agricultural consumers, since rural industrialization is still relatively insignificant, with the possible exception of China. Rural domestic consumption is much lower than urban consumption (see Table 14.4) for two main reasons:

The low monetary income of the rural populations prevents them from acquiring high-consumption equipment (air-conditioning, water heating, etc.).
Information about the possible uses of electricity apart from lighting remains extremely limited.

In the developing countries taken as a whole, lighting, a very small consumer, remains by far the primary use of domestic electricity in the rural environment. In countries that have significantly advanced the electrification of the rural populations, average domestic consumption (less distorted by affluent villagers who are big consumers) is stabilized at a very low level. In Malaysia, for example, 64% of rural households consume less than 600 kWh/year; 65% in Algeria consume less than 400 kWh; in the Philippines, 90% of the rural households consume less than 420 kWh/year; and an Indian survey indicates consumptions of 240 kWh/year.

Simultaneously, the constraints concerning the mechanical strength of the lines impose the oversizing of the conductor crosssection in relation to the power demand. The situation is further aggravated in certain cases by the fact that the sizing of the system is carried out by experts from the industrial countries according to their own consumption standards.

Hence the rural grids are often considerably oversized for the amount of energy they convey. Added to this is the fact that the demands are highly concentrated in time, creating a high peakload in the evening but with virtually no demand during the rest of the day. This unsound use of the peak leads to load factors that are much lower than those found in the cities: 20% in the rural environment in Salvador against 50% in the urban zones, and 10 and 40% respectively, in India.

The rural kWh cost is particularly sensitive to this parameter of "consumption level;" as shown by Table 14.5 (compiled from an average on Indian projects in the

Table 14.4
Electricity Consumption in Rural/Urban Areas (per user in kWh/year)

	Countries	Rural	Urban
Average	Thailand	200	4,000
	El Salvador	1,000	4,000
Domestic	Indonesia	600	2,000
	Brazil	600	2,500
	Tunisia	440	825
	India	200	not available

states of Maharashtra and Andhra Pradesh). Higher consumption per km of line leads to a much lower kWh unit cost due to better use of the investments.

Consumption figures are higher for programs intended for irrigation, but the mean demand set up remains very low in comparison with the capacity installed to cope with peakloads (this factor of the seasonal peakload in demand clearly appears in Figure 14.2). The problem of amortizing fixed costs thus remains unsolved. Hence the rural LV kWh cost is substantially higher than the urban kWh cost, due to the much higher capital costs for the same number of customers and due to consumption that is insufficient and/or irregularly distributed to amortize these fixed costs.

The need for overall national fairness and the rural development objectives nevertheless demand a solution to ensure that the rural population will obtain electricity at acceptable cost, giving rise to the principle of tariff equalization (total or partial), which helps to restore the balance between town and countryside.

However, this customary solution presents two drawbacks. On the one hand, it tends to mask the real cost of electricity in the rural areas, with the risk of directing the consumer toward decisions that are uneconomic for the country, and on the other, it could, in the medium term, lead to a financial deadlock. Hitherto, in fact, only the developed countries have achieved rural electrification rates approaching 100%. The deficit generated by the small number of rural customers, which is also non-negligible, is acceptable due to the scale of urban consumption. But one may well ask what would happen when a country like India, which is 80% rural, will be totally electrified, as the financial losses of the Indian electrical utilities due to rural electrification already amounted to $185 million in 1976/1977, largely nullifying the surplus of $64 million achieved from the rest of the operation.

There is a need for an overall consideration of the implicit aid that a country devotes to rural development and the way it is used to ensure that this aid is no longer focused on a single technological system, but used in accordance with the density of the area concerned, and its distance from the grid, with maximum effectiveness within the framework of current financial constraints.

Table 14.5
Sensitivity of kWh Cost to Consumption Level

Villages	Lines (km) MV	LV	Consumption (kWh/year)	kWh/km (LV + MV)	Cost/kWh (US$)
Lingi	5.1	2.0	4,830	680	1.03
Lachampur	3.4	2.5	5,780	980	0.82
Beempur	5.6	1.9	7,650	1,020	0.71
Sanghi	5.4	3.5	6,720	760	0.69
Dhanora	4.5	4.5	10,260	1,410	0.63
Jessaval	2.7	6.0	77,000	8,850	0.13
Mainpur	1.7	1.6	34,090	10,330	0.13
Sherpur	8.3	34.4	286,420	6,760	0.14
Kirzabed	1.8	2.0	70,720	18,610	0.09

Source: Estimates from World Bank tables.

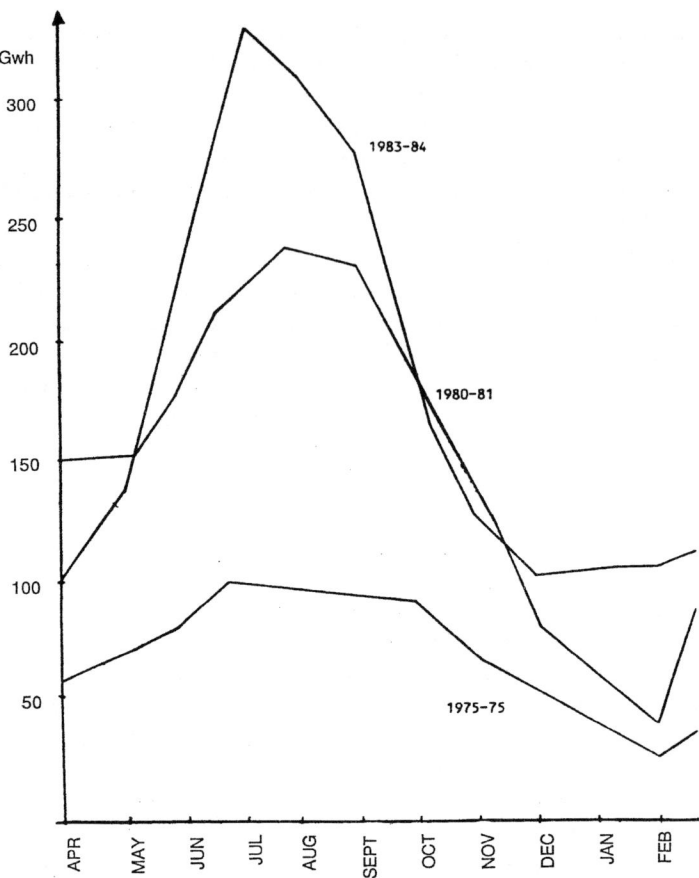

Figure 14.2 Consumption of electricity in the agricultural sector—Punjab, India. (*Source:* "Role of electricity in energy economy of India," S.K. Aggarwal)

Table 14.6
Examples of Subsidies to Agricultural Customers in Different Indian States
(in 1979 US cents)

State	Year	Mean LV kWh cost (US cents)	Agricultural kWh selling price (cts)
Andhra Prad.	1978/79	6.20	2.02
Assam	"	6.20	2.65
Bihar	"	6.40	0.88
Gujarat	1976/77	4.16	2.90
Karnataka	"	3.53	2.65
Punjab	"	4.67	1.89
West Bengal	"	6.20	4.41

Source: Energy policy, September 1982.

CONCLUSIONS: THE FUTURE OF GRID EXTENSION

The analysis of the characteristics of rural electrification in grids shown above reveals that the costs are particularly sensitive to several factors:

- distance
- population density (Table 14.2)
- consumption level (Table 14.5)
- shape of the load curve (Fig. 14.2)

In fact, the farther electrification spreads into the rural areas of the developing countries, the greater the distance and the lower the consumption density with pronounced peakload factors. This results in higher connection and kWh costs for new customers.

In the absence of a very strong political determination, backed by large funding resources, economic reason points clearly to the following scheme for the extension of interconnected grids: equipment of the large urban centers, followed by secondary centers, and finally the nonprofitable small villages and the remote hamlets (shown in Table 14.7 for India). This means that it is impossible for the countries initiating their rural electrification programs today to serve remote, low-population density regions in the medium term, unless they consider the use of stand-alone systems.

To conclude, let us recall two significant figures:

In the 1982/1992 decade, the capital requirements of the electrical sector in the developing countries, calculated by the World Bank, account for 72% of all external loans.

The fourth Malaysian five-year plan calls for the investment of $260 million in rural electrification over the 1981/1985 period, or 25% of the energy investments of the plan, merely to raise the country's rural electrification rate by 4%.

These figures seem clearly to pose the two central questions concerning rural electrification: to what extent is electrification a sound answer to the energy problems

Table 14.7
Rate of Village Electrification in India (Andhra Pradesh)

Population (inhabitants)	Total number of villages	Number of villages electrified	% of villages electrified
10,000	80	80	100
5–10,000	725	669	92
2–5,000	4,832	3,859	80
1–2,000	6,411	3,866	60
500–1,000	5,438	1,967	36
500	9,755	917	9

Source: Andhra Pradesh State Electricity Board.

of the rural population of the Third World countries? What the the respective roles of grid extension and decentralized electrification in areas where electrification has an important role to play? It appears that the development of rural electrification exclusively by the extension by interconnected electrical constitutes a risky policy in many respects:

Financial risk. A highly capital-intensive technique, investments that are essentially less divisible, and the uncertain if not negative profitability of rural electrification projects accentuate the precariousness of the financial balance of the electrical utilities.

Economic risk. The completion of rural electrification programs (rates higher than 70%) will take place, as we have shown, at sharply increasing cost. In a context of capital scarcity and rising foreign debt, hence of the limitation of investments, this could mean the discarding of other programs (agriculture, road infrastructure, etc.).

Technological risk. The weight of habit, and the existence of a technical potential, often lead to the preference of a particular technology, even if it is too costly. This "technological forcing" could prove counterproductive and could create burdens that would considerably hamper the possibility of a solution by other systems that may be better adapted to the utilization of local resources.

Social risk. Very few countries among the least advanced can seriously consider meaningful rural electrification (exceeding 50% of the rural population) only through the extension of interconnected grids. For countries with intermediate income, these rates may reach 70 to 80%, but in any case, an answer must be found to the energy problem of the unconnected populations (remote and sparsely populated areas) to avoid their permanent marginalization.

15

Decentralized Electricity Generation by Internal Combustion Engines

---

DESCRIPTION

Internal combustion engines divide into two main categories: compression ignition (more commonly known as diesel engines) and spark ignition, which generally run on petrol (gasoline) or kerosene.

Manufacturers rate engines at their maximum power, but in practice engines are derated to run at somewhat lower power levels. Diesel generating sets are available in power ratings from about 1.5 kW up to tens of megawatts, whereas spark-ignition (gasoline, kerosene, or l.p.g. fueled) generating sets can be obtained with outputs as low as 500 W. It is unusual to use spark-ignition engines for generating set outputs over about 5 kW since they are less efficient and less robust than their diesel equivalents; but their simplicity and low cost make them the *only* choice for loads of about 1 kW or less, and they are attractive for higher power ratings when portability or low first cost is important.

Internal combustion engines tend to be badly matched to the needs of small, continuous power applications of under 1 kW, which are common in remote areas off the grid. Therefore, it is normal to supply smaller, continuous electrical loads when they are important enough to justify the cost (e.g., telecommunications) by using a diesel generating set (usually with a backup/standby unit) to charge a battery bank, which in turn acts as a power conditioner by storing electrical energy and releasing it at a lower power level than can be conveniently supplied directly from an engine.

STATE-OF-THE-ART

The internal combustion engine is the world's most common prime mover and it has had more than a century of intensive development. This is a mature technology, which is mass produced with only limited scope for improvement. However, it is often incorrectly applied and economies may be found through better system sizing and improved operating techniques.

Engines are constrained by a general dependence on petroleum fuels. There are, however, possibilities (mostly under development) to use nonpetroleum fuels based on biomass. The primary options include biogas, alcohol, producer gas (from wood or charcoal), treated vegetable oils, certain plant latex, saps, and resins. Most biomass fuels for internal combustion engines at present still either cause technical problems or they do not offer any significant economic advantage over present-day petroleum fuel costs, but it is expected that in time biomass fuels will become more economically attractive.

Diesel engines are inherently more expensive to manufacture, but they compensate by being more efficient. They also are more reliable and long-lasting.

Small light weight (low cost) engines tend to have short useful lives, because a high power/weight ratio is normally achieved by running an engine at high speed and wear in machinery is a function of speed. For example, an engine with cylinders of 30 mm diameter (i.e., perhaps 3 kW power rating) usually has a useful life of only about 1,000 hours between major overhauls or complete replacement, whereas large diesels with cylinders of around 150 mm diameter would typically achieve over 10,000 hours before sufficient wear has taken place to require a major overhaul.

Small engines are usually derated to about 70–80% of their rated power; e.g., a 5 kW (rated) engine will be necessary to produce a continuous power output of 3.5–4.0 kW at its shaft. Because there are losses in a generator (a small unit may only be 80% efficient), the electrical output from such an engine when used in a generating set would be 80% of 3.5–4.0 kW or 2.8–3.2 kW. Excessive derating is to be avoided, as (particularly with diesels) running at a fraction of the design power tends to cause carbonization and gummed-up injectors. Also, the engine efficiency will be poorer than normal. Spark-ignition engines suffer more from reduced efficiency at part-load; typically, their specific fuel consumption increases by 20% at half-load and by 60% at quarter of their rated power.

Unfortunately, it is easy to run an inefficient engine system without realizing it, because any shortfall in performance is easily made up by running the engine at a higher throttle setting or for longer than would otherwise be necessary. Gasoline engines are capable of being 25–30% efficient, whereas diesels are 30–40% efficient, but this applies to an optimally tuned engine running minus most of its accessories. In reality such figures are optimistic. The difference between theory and reality is the worst with the smallest sizes of engines, which are inherently less efficient than larger ones (the above figures apply only to engines of over 5–10 kW rating). In practice, engines are generally not in perfect condition, and they are run at nonoptimum speed with nonoptimum loads.

The smallest engines of around 1 kW raing can be as poor as only 10% efficient. Small diesels (the smallest are generally about 1.5 to 2 kW) will probably be better than 25% efficient as engines, but components like the injection pump, cooling fan, and water circulating pump (all parasitic energy consumers) can reduce the fuel-to-useful-shaft-power efficiency to under 15%, because these accessories take proportionately more power from smaller engines. When all the possible losses are compounded they yield a theoretical worst total efficiency as low as 0.5%, whereas, the best factors compounded together give 27%. This optimistic figure is only even theoretically feasible when every single possible source of losses has been optimized for the duration of each and every duty cycle (which, of course, is impossible in practice). Therefore, in reality most small engine powered systems (under 10 kW) only achieve 5 to 15% total efficiency, but larger diesel-generating sets with well-matched loads might achieve better than 20%.

SPECIFIC CONSTRAINTS

Petroleum-fueled engines differ from most of the renewable energy alternatives in having low first costs and high recurrent costs. First costs are subject to factors such as discounting by manufacturers, which result in great variation of sales price for some models of generating set procured from different sources (middlemen). Recurrent costs are influenced primarily by fuel costs and consumption rates plus O&M costs.

Fuel costs are easily determined, but rates of consumption are highly variable, especially with small engine systems. Manufacturers' brochures are not helpful in giving guidance; many for the smaller sizes of engine give no figures at all for fuel consumption and those that do generally present the information in a not easily interpreted form. Even when clear fuel consumption figures are stated, they invariably are given for full-rated load conditions and there is no indication of what performance might be expected under more typical part full-load conditions.

Engine powered systems requires significant human intervention both for regular supervision (checking, refueling, starting, stopping) as well as more specialized technical support for regular maintenance and repairs. Such support tends mainly to be available on or near main roads and provincial towns due to the prevalence of motor vehicles that have similar needs, and is absent off the main roads where most of the rural population lives. Hence there are high operational overheads associated with installing engines off main transport routes.

A major constraint is fuel shortages and distribution problems common in many of the poorer countries. These are due both to problems of physical distribution (transport to remote areas) and financial problems (lack of foreign exchange). A similar constraint relates to the provision of lubricating oil. This is primarily why renewable energy systems running independently of an imported fuel supply are of interest for such areas.

Engine-powered systems generally rely almost entirely on imported fuel and imported spares and components; therefore they represent a contribution to the chronic balance of payments and debt problems of many of the poorer countries. Aid monies are often available to assist with capital investment, but less commonly to meet recurrent costs. Investment of aid monies in engines commonly leads to financial problems for the receiving agency in sustaining the resulting recurrent costs.

COSTS

Table 15.1 indicates that whereas a 1 kW gasoline generating set may typically involve amortization (capital cost)/Life at $0.067/kWh, a 100 kW diesel unit having a first cost 50 times as great will actually only amortize at $0.02/kWh. This is because of the much higher productivity and longer life of the larger system, assuming, of course, than an adequate load to absorb 50% of its potential capacity actually exists.

Recurrent costs are made up of three components:

fuel costs
maintenance and repair costs
attendance costs

Parameters affecting unit fuel costs include the oil company price, plus transport and storage costs, plus an allowance for any losses. The system efficiency dictates how effectively the fuel is converted into output energy.

Figure 15.1 illustrates how the fuel component costs are dominant both for gasoline-generating sets (which are relatively inefficient) and for larger diesels, where the capital cost component is small due to more favorable costs per kW of installed capacity.

Maintenance and repair costs are very variable from one situation to another. The figures used in Table 15.1 assume "ideal" maintenance. In practice, careless or inadequate maintenance can have a profound impact both on total costs and on engine life.

Attendance costs are also generally involved. Every time an engine is started, it is usually necessary to check the oil (and coolant); sometimes fuel, oil, or coolant needs replenishment. These routine requirements demand regular attendance. Also it is necessary for someone to be on hand while an engine is running to react to any irregularities or problems that may occasionally occur. Engines can be fully automated with remote-controlled starting and stopping (this is sometimes invoked for telecommunications power equipment), but the cost of the various control devices is very high and they themselves demand specialist occasional maintenance.

POTENTIAL FOR LARGE-SCALE DEVELOPMENT

Internal combustion engines are already highly developed and widely deployed. The main new potential is for quieter and more economical engines with reduced exhaust emissions. There is also potential for the development of accessories to improve their cost effectiveness, such as heat exchangers to allow waste heat to be used for supplementary purposes such as water heating.

Table 15.1
Generating Set Running Costs

Rating (kW)	Gasoline			Diesel				
	1	5	2	5	10	20	30	100
Capitol cost ($/kW)	500	300	1,200	900	700	500	400	250
Installed cost ($)	500	1,500	2,400	4,500	7,000	100,000	12,000	25,000
Overall efficiency (%)	4	8	12	15	18	22	28	32
Operating life (hrs)	1,500	4,000	5,000	10,000	10,000	15,000	20,000	30,000
Output during life (kWh)*	750	10,000	5,000	25,000	50,000	150,000	300,000	1,500,000
Cap. cost/life (c/kWh)	87	15	48	18	14	7	4	2
Fuel cost (c/litre)	60	55	45	45	45	45	40	35
Fuel consumption (l/life)*	2,083	13,889	3,968	15,873	26,455	64,935	102,041	46,429
Fuel costs ($/life)	1,250	7,639	1,786	7,143	11,905	29,221	40,816	156,250
Fuel component (c/kWh)*	167	76	36	29	24	19	14	10
O&M costs ($/100 hrs)	50	50	80	80	80	80	80	80
O&M component (c/kWh)	10	2	8	3.2	1.6	0.8	.533	.16
Net unit cost (c/kWh)	243	93	92	50	39	27	18	12
Operating hours/day	2	4	6	8	20	10	10	10
Absolute life (years)	2.5	2.74	2.28	3.42	2.74	4.11	5.48	8.22
Number of starts (no/life)	750	1,000	833	1,250	1,000	1,500	2,000	3,000
Attendance cost (c/kWh)†	100	10	17	5	2	1	0.67	0.20
Daily output (kWh/day)	1	10	6	20	50	100	150	500
Gross unit cost (c/kWh)	343	103	108	55	41	28	19	12

*Assuming average useful output is 0.5 of rated power.
†Assuming attendance costs US$1.00 per start.

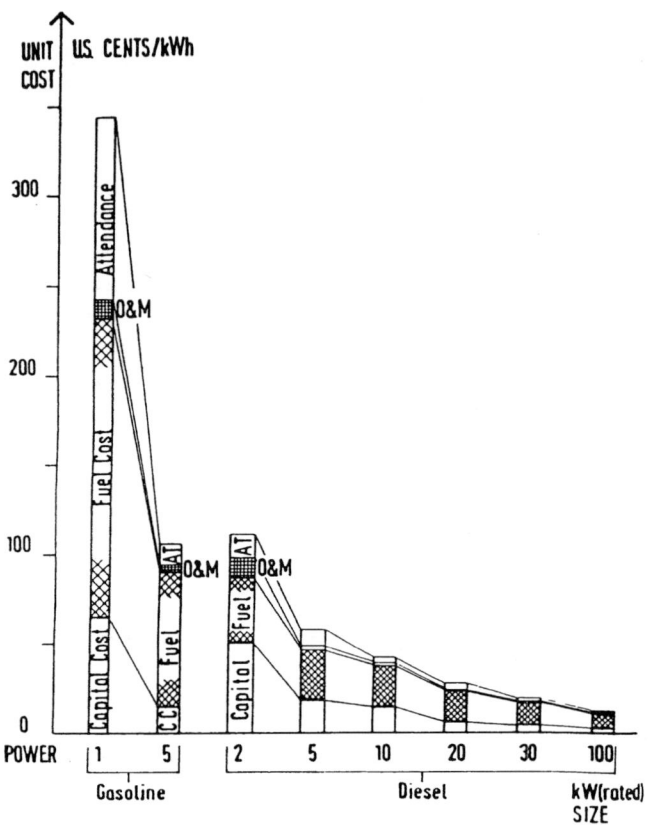

Figure 15.1 Unit costs of various sizes of gasoline and diesel generating sets, assuming that average output = 0.5 of rated power and attendance cost = $1.00 per start.

ADVANTAGES AND DISADVANTAGES

The main advantages of internal combustion engines are:

1. Very flexible power source: can be started rapidly on demand and can cope easily with sudden variations in demand.
2. High power-weight ratio makes it easily portable.
3. Low capital cost (needs least investment to obtain installed capacity).
4. Widely available and widely understood and supported.

The main disadvantages are:

1. Generally dependent on petroleum-based fuels (alternative fuels still present problems or are not as cost-effective).

2. Needs significant human interventions for operational and maintenance functions.
3. Inefficient with particularly high unit operating costs for small-scale power applications: difficult to match with sub-kW power applications; difficult to match with sub-kW power demands without either the use of batteries or with accepting very poor system efficiency and very high unit costs.
4. High recurrent costs generally.
5. Vulnerable to serious damage as a result of poor maintenance or poor supervision.

CONCLUSIONS

Although internal combustion engines are the de facto standard for remote power generation, there is scope for improving the efficiency with which they are applied and possibly for simplifying maintenance functions.

Alternative biomass-based fuels for internal combustion engines are potentially important, but further development is needed before they will become both economic and practical for really widespread use.

Internal combustion engines are least effective for subkilowatt applications, and it is here that they are likely to be replaced by solar or wind-powered renewable energy systems.

16

Electricity Generation by Micro-Hydro Power Stations

GENERAL DEFINITION/SIZES

There is no precise definition of "micro-hydro," but a commonly used definition includes all systems up to 500 kW. The minimum size of system commonly used is around 5–10 kW, although some are as small as 1 kW.

Qualitatively, micro-hydro generation implies a localized, rural system without extensive transmission lines or voltage transformation, although similar hardware may also be used for grid-feeding as well as for localized generation. Micro-hydro also implies a "run-of-river" system in which no attempt is made to impound or store water with a dam or earthworks, so the system is subject to seasonal fluctuations of river flow and can never use more water than the full flow of the river at any given time.

Micro-hydro electric systems are generally rated for peakhead and flow conditions. Because flow rates vary seasonally, many systems are therefore capable of achieving their rated power only at certain times of the year.

STATE-OF-THE-ART

Micro-hydro systems in the power range of interest may use heads from as little as about 2 m up to 500 m and flow rates of a few liters/second up to 20 m3/s as indicated in Figure 16.1. The main application for micro-hydro turbines is electricity generation, although in China, for example, mechanical postharvest processes like rice milling or oil expelling produce a better financial return than pure electricity generation, and some systems are therefore dual purpose.

China is by far the largest user of micro-hydro, with approximately 90,000 installations of less than 5 MW having a total rated capacity of 6,600 MW. As a result China has a uniquely well developed micro-hydro, which is probably more than 90% of the world total industry. Other regions, particularly Europe and the United States,

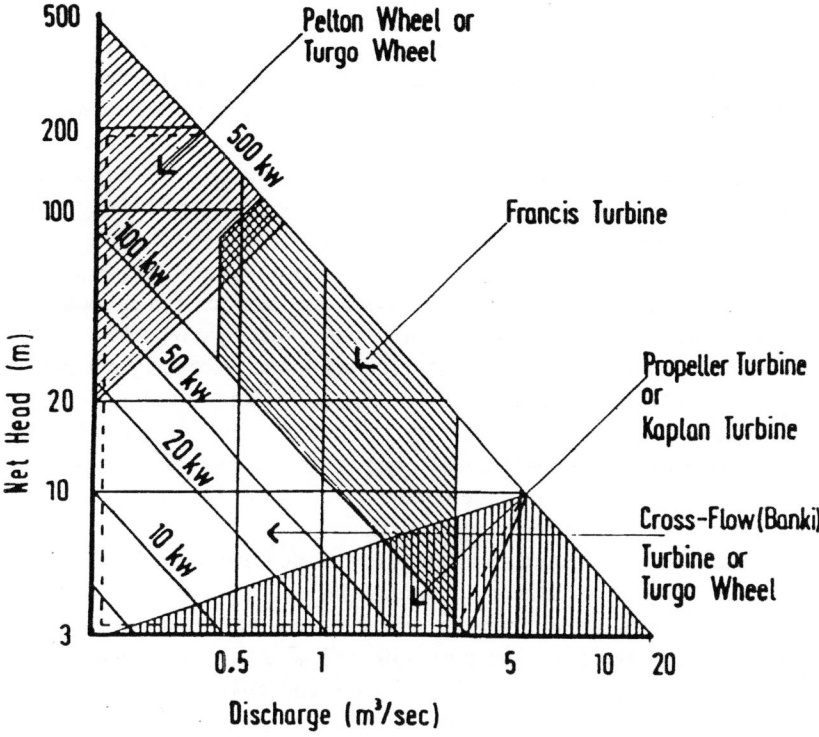

Figure 16.1 Application ranges of micro-hydro turbines.

also have micro-hydro industries but much smaller home markets. Many western manufacturers are survivors from an earlier era (pre-1930s). Before the advent of rural electrification, there was a larger home market, but there has been a recent revival of interest in micro-hydro, particularly in Scandinavia, Canada, and the United States with numerous new products appearing. Nepal, Pakistan, Sri Lanka, Thailand, Colombia, and Peru are other developing countries that are now manufacturing and installing micro-hydro energy sources.

Different types of turbines are necessary for use at different heads and flows. Low head turbines achieve a relatively high rotational speed in relation to the water velocity passing through them but would turn too fast if used at high heads. Most low head turbines are based on a propeller-type runner and are known as reaction turbines since the casing runs full of water and rotation is caused by the reaction of the propeller blades on deflecting the path of the water. At medium heads Francis or Banki turbines are used (the latter is less efficient but more easily manufactured). At higher heads, impulse turbines (where water from a nozzle strikes deflectors on the turbine rotor) such as the Pelton wheel come into their own (see Figure 16.1). Pelton wheels are particularly efficient over a wide range of flows, but they only turn fast enough for electricity generation when used on high heads.

There are no hard and fast rules on the ranges of heads appropriate to different types of turbine. For example, small low-powered Pelton wheels can be used at much lower heads than larger units because their smaller diameter means they turn faster. Similarly, large propeller turbines are able to be used at higher heads than small ones for the reverse reason.

Turbine technology has not changed very much since the 1930s, but the introduction of power electronics has allowed some important innovations to take place during the 1970s and 1980s in the field of control. Traditional turbines were regulated by varying the flow through them in response to control signals from a mechanical speed sensing governor. Such systems, especially on a small scale, tend to be inefficient, inaccurate, not wholly reliable, and expensive in relation to the costs of the complete system. Therefore small systems hitherto produced inadequately regulated outputs or tended to be too expensive (or both). Recently it has become possible to regulate the speed of a micro-hydro system by electronically switching in or out extra ballast loads. This technique is known as electronic load-control and can be done with greater accuracy and reliability at much reduced cost. This has greatly improved the economics of micro-hydro systems.

There have also been technical developments in low head systems. Low head is much more difficult to implement economically than high head, since the flow rate of water passing through the turbine is inversely related to head; the lower the head the more water and hence the larger the turbine has to be. This problem has led to the development of a variety of novel low-head propeller-type turbines; notably bulb turbines (where the generator is mounted at the turbine hub submerged in the flow), rim turbines, right-angle drive turbines, and others. Submersible electric pump technology has led to the development of less expensive bulb-turbines with totally sealed and submerged generators based on pump designs.

CONSTRAINTS TO WIDESPREAD DIFFUSION

The primary constraint is locating sites with adequate flow and heads to enable low cost and simple installation. Low head sites are statistically much more common than higher head ones and they also tend to be nearer to centers of population that can use the electrical power generated.

The flow rates of many rivers vary so much seasonally that all the year around micro-hydro exploitation is impossible or noneconomic. Sometimes seasonal flooding can also cause problems by damaging, flooding, or silting up the system. Many of the Chinese installations mentioned earlier are on river systems that are extensively controlled and canalized and that therefore do not suffer extremes of flow of uncontrolled rivers.

Even in areas with obvious micro-hydro potential, there is often a lack of reliable data on river flows and their seasonal variations, which makes design and planning of systems problematic. Nevertheless, even without rain gauge and stream flow gauging data, good hydrological extrapolations can often be made. For example, a study by the Institute of Hydrology in the U.K. developed a simple relationship to predict flow duration curves for southwest Sri Lanka from the mean annual rainfall. A further important constraint, which is often overlooked by planners, is that even if an ideal site exists, it is vital that there is a viable economic application for the electricity. Otherwise a very low load factor will be realized and the system will fail to pay for itself. Where an application does not exist, it is important to consider developing one in conjunction with any micro-hydro installation; it is pointless to produce power if noone makes effective use of it.

A major constraint with western-style micro-hydro is very high engineering overheads because all too often the approach is to design installations similar to large-scale systems, that is, almost every installation is treated as a unique, one-off, multidisciplinary engineering design project and hence incurs high design costs. The Chinese have sought to standardize on plant and machinery and also on installation configurations. This is so that the best "standard" installation for a new site can be selected and then copied without the need for original customized design work. In other words they have introduced a kind of "cookbook" approach where small hydro can be implemented using nonspecialized engineering skills by following a standard "recipe" and procuring standard equipment. This appears to be an important means to overcome the high overheads that often make western micro-hydro appear uneconomic. Lack of local engineering and management capability combined with lack of familiarity with micro-hydro also is a common constraint in developing countries. Even with the Chinese "cookbook" approach, it is essential to have a certain basic local engineering capability.

COSTS

The main parameters that affect the economics of a micro-hydro electric system are its capital cost and the load factor achieved. The load factor is the proportion of electricity capable of being generated that actually can be sold to a user and yield revenue. On average, demand will be lower than the system is capable of delivering, so the load factor approximates to the ratio of average load to peak load.

Efficiency is both high (compared with other prime movers) and fairly constant for all types of hydro-turbines (usually 70–85%). Also, O&M costs are relatively small. If a micro-hydro system is installed in an area where previously electricity has never been used (so initially there are no appliances or equipment to provide a load), it can prove difficult to develop an adequate load factor, since lighting, which is the first thing people demand, uses little power, yields little revenue, and is only needed for a few hours a day.

Capital costs divide into three main categories, machinery or equipment costs, civil engineering costs, and management/engineering overheads. All are quite variable depending on local site conditions, labor costs, and financial factors. Total installation costs can be in the range $1,000 to $10,000/kW of installed capacity, (around $1,000–2,000/kW is typically a common level for an economically attractive micro-hydro system). Of this, machinery and equipment costs will normally be in the region of $200–$1,000/kW of installed capacity.

The main parameters influencing unit costs are the capital cost (and financial parameters used to amortize it) and the load factor (i.e., ratio of mean load to rated or peak load). The benefits consist of the gross energy that could in theory be produced, multiplied by the load factor.

The total costs are primarily the annualized capital cost, since O&M costs for modern systems are very low (typically 1–3% of capital cost) and the operational life of most such systems runs to decades. Electronically controlled systems can run completely unattended and are usually subject only to a weekly inspection. Experience in Sri Lanka indicates that delivered electricity costs in the range from $0.09–0.34/kWh, being inversely proportional to load factor. Realistic load factors are in the range 0.1 to 0.4.

POTENTIAL FOR LARGE-SCALE DEVELOPMENT

Micro-hydro is limited to areas with adequate river runoff (i.e., generally hillier regions having adequate rainfall), and reasonably constant flow through all seasons. Although this limits the areas where micro-hydro may be used, few regions have exploited this resource to any great extent so that only a limited proportion (probably only a few percent) of the world's micro-hydro potential is currently being used. Even China only uses an estimated 43% of its exploitable hydro resource.

ADVANTAGES AND DISADVANTAGES

The main advantages of micro-hydro power are:

1. It is a relatively concentrated energy resource.
2. The mechanisms for converting it to electricity are well developed and available.
3. The conversion process is efficient (ca. 80%).
4. The technology is robust, long-lasting, and requires little human intervention.
5. Under favorable circumstances hydro represents one of the least costly methods of electricity generation.

The main disadvantages are:

1. Suitable sites are not common, although the majority are not being exploited at present.
2. Suitable sites tend to be absent in populated areas.
3. There is always a maximum power output possible from any given site, which limits further expansion.

4. The technology is often not well understood, which leads to failure to use it in appropriate situations or sometimes to the incorrect use of it.
5. It is necessary to develop an adequate load factor, which requires that a consumer network often needs to be created in areas not previously having electricity.

CONCLUSIONS

Micro-hydro electric power, when properly implemented in areas with adequate hydro resources, can offer one of the least costly means for generating electricity. Moreover the equipment has been developed, is available, and is robust, long-lasting, and needs little technical support once installed.

There is a widespread need to determine micro-hydro potential by evaluating river flow data that is available and filling gaps in the knowledge by a program of stream gauging or by analysis of catchment areas and rainfall data.

It is important that installations are standardized and installed by formula rather than as individually designed and engineered projects, as a major factor in the costs of many micro-hydro projects are design and planning overheads. It is essential that remote micro-hydro projects are provided with economically viable electrical loads; this should be considered at the planning stage.

17

Electricity Generation by Photovoltaic Power Stations

This chapter deals with medium capacity (10–200 kW) solar photovoltaic power stations converting solar energy into 220V (or 120V) alternating current by means of dc/ac inverter.

STATE-OF-THE-ART

Medium capacity photovoltaic power stations are still in the pilot stage, with less than 30 plants throughout the world. The earliest projects in the United States and France are less than 10 years old.

A photovoltaic power station comprises:

- the photovoltaic module field: about 12 m^2 per kW peak installed capacity
- battery bank
- cabinets containing the charge and discharge regulators
- dc/ac inverter

Two types of photovoltaic power station can be distinguished: stand-alone power stations not connected to a grid, which require storage by batteries, and power stations operating usually with "solar tracking" but without built-in storage (currently under development in California) coupled to an interconnected grid. Only the former is presently applicable to developing countries.

The technology applied is relatively complex and delicate, requiring qualified personnel for the maintenance of the electrical and electronic parts. Present installations show good reliability on the whole, with the possible exception of the inverters. Major progress can be expected in inverter reliability, and the main uncertainty concerns the lifecycle of the batteries (about 7 years).

COSTS

Medium capacity solar photovoltaic power stations are expensive and do not permit rapid development in the short term. For example, a 50kW solar power station typically generating 150 kWh/day in a sunny tropical country costs about $700,000 resulting in a kWh cost of about $1.80. (This typically requires 5kWh/m2/day of solar radiation falling on the plane of the array. However, the array output is likely to be 75% of rated power under standard conditions due to degradation, temperature

Table 17.1

Cost of kWh Produced by Medium Capacity Photovoltaic Power Stations, KAW PV Power Station, French Guyana*

		Interest rate in %		
		5	10	15
Installed capacity W	35000			
Energy output kWh/day	75			
Capital cost in F	5000000			
Capital cost in US$	667000			
Cost per watt peak in F	143			
Cost per watt peak in US$	19			
% Modules in capital cost	44			
% Batteries in capital cost	10			
% Conv/reg. in capital cost	8			
% Other in capital cost	38			
Depreciation period modules: years	20			
Depreciation period batteries: years	7			
Depreciation period conv/reg: years	15			
Depreciation period other: years	15			
Labor + overhead in % cap. cost	1			
Maintenance cost in % cap. cost	2			
Calculated annual costs				
Module depreciation		176534	258411	351475
Battery depreciation		86410	102703	120180
Conv/reg. depreciation		38537	52590	69407
Other depreciation		183050	249800	324932
Total depreciation F/year		484531	663504	864995
Labor + overhead F/year		50000	50000	50000
Maintenance F/year		100000	100000	100000
Total annual costs F		634531	813504	1014995
Total annual costs US$		84600	108500	135300
kWh cost ex plant F/kWh		23	30	37
kWh cost ex plant $/kWh		3	4	5

*The capital cost includes over 30% engineering costs. Assuming battery life of 5 years, the kWh cost rises to 31 FF (for i = 10%). For module life reduced to 15 years, kWh cost increases by 1 FF. The high maintenance costs include very difficult access. In such a site, the diesel kWh cost would be about 10 F/kWh. Values are rounded off. 1 US$ = 7.5 FF.

derating–cells are often 25–30° C above the standard 25° C—and uncoupling of array when battery is fully charged and there is no load. Finally system output is about 80% of array output due to losses in battery charging and in inverter, etc. Hence actual output 3kWh/h2/day, i.e., 3kWh/day per peak kW.)

Tables 17.1 and 17.2 give a number of analyses of kWh costs based on recent examples, including PV power stations in Pakistan and the Aghia Roumeli power station in Greece. Table 17.2 gives the breakdown used in Table 17.1 based on an

Table 17.2

Cost of kWh Produced by Medium Capacity Photovoltaic Power Stations, Aghia Roumeli PV Power Station, Greece, EEC Project*

		Interest rate in %		
		5	10	15
Installed capacity W	50000			
Energy output kWh/day	150			
Capital cost in F	7500000			
Capital cost in US$	1000000			
Cost per watt peak in F	150			
Cost per watt peak in US$	20			
% Modules in capital cost	40			
% Batteries in capital cost	15			
% Conv/reg. in capital cost	15			
% Other in capital cost	30			
Depreciation period modules: year	20			
Depreciation period batteries: years	7			
Depreciation period conv/reg: years	15			
Depreciation period other: years	15			
Labor + overhead in % cap. cost	1			
Maintenance cost in % cap. cost	1			
Calculated annual costs				
Module depreciation		240728	352379	479284
Battery depreciation		194422	231081	270405
Conv/reg. depreciation		108385	147908	192394
Other depreciation		216770	295816	384788
Total depreciation F/year		760305	1027184	1326872
Labor + overhead F/year		75000	75000	75000
Maintenance F/year		75000	75000	75000
Total annual costs F		910305	1177184	1476272
Total annual costs US$		121375	156960	196920
kWh cost ex plant F/kWh		17	22	27
kWh cost ex plant $/kWh		2	3	4

*The capital costs include high engineering costs, given the pilot character of this EEC project. Life cycles are considered in a favorable context. The kWh cost obtained is not at all competitive with the diesel kWh cost observed locally, which is less than 3 FF. Values are rounded off. 1 US$ = 7.5 FF.

Table 17.3

Cost of kWh Produced by Medium Capacity Photovoltaic Power Stations, Pakistan Village Project (UNDP/DTCD) Assisted)

		Discount rate in %		
		5	10	15
Installed capacity W	20000			
Energy output kWh/day	60			
Capital cost in US$	280000			
Cost per peak watt US$	14			
% Modules in capital cost	46			
% Batteries in capital cost	18			
% Inverter/regulators, etc.	13			
% Other (civil works + installation and training cif)	23			
Depreciation period modules: years	20			
Depreciation period batteries: years	7			
Depreciation period inv/reg: years	15			
Depreciation period other: years	15			
Labor + overhead in % cap. cost	1			
Maintenance cost in % cap. cost	1			
Calculated annual costs				
Module depreciation $		10300	15100	20600
Battery depreciation $		8700	10300	12300
Inv/reg depreciation $		3500	4800	6200
Other depreciation $		6200	8500	11000
Total depreciation $/y		28700	38700	50100
Labor and overhead $/y		2800	2800	2800
Maintenance $/y		2800	2800	2800
Total annual costs		34300	44300	55700
kWh cost ex plant $/kWh		1.57	2.02	2.54

analysis of costs for two villages in Pakistan in 1983. Costs are an average for the winning bid and the next lowest bid for each village.

The medium-term prospects for a very sharp cut in costs are limited, because the solar cells only count for 30 to 45% of capital costs. The other components (batteries, inverters, etc.) are fully developed products for which no rapid price decrease can be anticipated. The scale effect is likely to remain modest because the series are too small.

COMPETITIVENESS WITH OTHER ALTERNATIVES

Medium power photovoltaic power stations for village electrification are not competitive today by comparison with diesel plants. Capital costs are nearly 20 times higher, and annual costs are themselves significant (battery renewal, maintenance by skilled personnel, often over long distances). They are justified in certain cases if the site concerned is isolated.

Operating as hybrid units with a diesel plant, their profitability is just as difficult to evaluate, since the PV plants must have a kWh cost that is lower than the cost of the diesel fuel saved (including, if applicable, savings in the maintenance costs of the diesel plant).

An analysis of costs of solar PV versus diesel generators for multiuse electric power undertaken by the Meridian Corporation for USAID and USDOE gives the sensitivity to capital costs, discount rates, diesel genset lifetime, fuel costs, and insolation. It concludes that the breakeven for PV is between 2 and 16 kWh/day over a wide range of assumptions, with 4 to 8 kWh/day being more typical. This means, in effect, that solar PV is competitive only with small diesel sets operating at part load, a situation that would rarely be found in village electrification.

ADVANTAGES AND DISADVANTAGES

Medium capacity PV power stations offer the following advantages:

1. Autonomy or reduction of fuel consumption.
2. Energy supplied to the local grid in 220 V ac enabling customers to use standard equipment.
3. Good adaption to initial demand because of their adjustable character.

They present the following disadvantages:

1. Excessively high capital costs.
2. Low reliability since the failure of one component could cause total shutdown.
3. The need to establish a distribution network, which is expensive in a sparsely populated area and which could be supplanted by small individual PV generators that are cheaper and more reliable.
4. Skilled personnel required for maintenance.
5. In case of shortage of energy (insufficient insolation for example), a backup diesel generator may be required.
6. The difficulty of adapting to a rapid growth in customer demand at low cost. The marginal cost of each additional kWh is very high in this case, in contrast to diesel.

FUTURE DEVELOPMENTS

The solar PV industry is currently in a state of flux. Current technology is based on solar cells each consisting of a wafer of very pure (electronic grade) monocrystal or cast polycrystal silicon about 10 cm in diameter and 0.2–0.4 mm thick. Each generates about 1 peak watt under standard conditions (1kW/m2 solar radiation and with so-called AMI spectral distribution and 25° C cell temperature) and has an efficiency for converting sunlight to electrical energy of 10–12%. Total world production is about 30 MW/year including an increasing percentage of the newer amorphous silicon cells, but single production lines do not exceed 4MW/year. Presently, prices of $5/W are quoted for large (multi-kW) orders. Even with production lines as large as 20–40 MW/year, it seems unlikely that prices could be reduced below $2/W with this technology. However, three new technologies are now being developed, each of which shows strong promise of achieving $1/W when produced in quantity.

1. Thin film technologies, primarily amorphous silicon, but also cadmium telluride (CdTe) and copper indium diselenide (CuInSe2). These have an efficiency of around 5% in current production but 10% efficiency is expected in the near future. It is also possible to combine two thin film materials having different spectral responses, one on top of the other, to form a tandem cell with efficiencies of 15% or higher.
2. Ribbon technologies in which typically 10-cm-wide ribbons of silicon about 0.1 mm thick are drawn from the melt. One version, dendritic web ribbon silicon, is believed to have the potential of meeting cost and performance targets and is already reaching 15% efficiencies under near production conditions.
3. Point contact silicon concentrator cells under intensive development bu the U.S. power utility industry. These consist of small (1-cm square) cells of superpure silicon with 70–80 thousand point contacts prepared by masking technologies developed for integrated circuits. These cells will be combined with low cost plastic fresnel lens concentrators (500X) and have demonstrated the highest conversion efficiencies so far achieved (27.5%).

Multimegawatt grid connected advanced PV systems should have installed costs of no more than $2/W including PV arrays plus so-called balance of system (inverters, installation, etc.). They will require no batteries and associated buildings, charge regulators, and so on. High efficiencies will result in reduced area related costs (land, shipping, installation). Sun-tracking will give 40% greater sunlight catchment (for equal peak kW rating) compared to static arrays (although concentrator systems will require direct sun). Together with avoidance of battery losses and higher efficiencies from large inverters, outputs per peak kW will be doubled to typically 6 kWh/day per peak kW. Finally, utilities operate on 30-year depreciation cycles, which should be attainable with some of the new technologies. On the above assumption and with operation and maintenance costs at 2% of capital costs, discount rates in the range of 5–15%, grid-connected solar PV, in suitable locations would provide power at costs in the range of US$0.06–0.16/kWh by the late 1990s (see Table 17.4). At these prices a very large market is expected and low cost PV modules should have an increasing impact for use as small energy sources, with or without battery storage or diesel backup.

Table 17.4

Cost of kWh Produced by Photovoltaic Power Source With and Without Batteries and
Inverter*

In certain applications, e.g., water pumping, small desalination plants, oil seed pressing, PV
power sources may not require battery storage (or require only minimal storage) and in some
cases may also not require ac inverters. This may effect substantial cost benefits not only
from the direct cost saving, but indirectly through savings on civil works, shipping costs and
installation costs, and also because of improved system efficiency. The following compares
costs of similar systems with and without batteries and inverter based on the assumption of
Table 17.3.

Capital cost $10,000/kW(peak) of which $6,000/kW is array and $4,000/kW is BOS.

		Discount rate %		
		5	10	15
Module depreciation ($6,000 20y)	$	480	700	960
Balance of system ($4,000 15y)	$	390	530	680
Labor and overhead (1% cap. cost)	$	100	100	100
Maintenance (1% cap. cost)	$	100	100	100
Total annual costs	$	1,070	1,430	1,840

*Annual production from 1kW (peak) at 5kWh/m²d on array is:
    4×365 kWh = 1460 kWh without battery,
    3×365 kWh = 1091 k with battery.

Cost $/kWh	$	0.73	0.98	1.26
Cost for full system from Table 17.3.	$/kWh	1.57	2.02	2.56

Costs are reduced by about 50% if storage and inverters are not required (most of the cost
saving comes from doing without batteries). Also useful output is 3kWh/d with batteries and
4kWh/d without batteries.

CONCLUSIONS

Medium capacity solar photovoltaic power stations are not economic today and in
the medium term, except in very specific situations. Although they are quite attractive in principle, they do not offer an ideal response to the needs of the developing
countries in their rural electrification policies. In the absence of a confirmed market,
their development remains limited to demonstration operations.

These conclusions only concern rural electrification. On the contrary, low and
medium capacity power stations offer excellent technical and economical alternatives
to supply professional units in remote areas, such as telecommunications relays,
transmitters, and so on. In the slightly longer run, the availability of low cost PV
modules, within 5 to 10 years, can be expected to have a major impact on both large
and small energy sources in developed and developing countries.

18

Wind Electricity Generators

GENERAL DEFINITION/SIZES

Wind electric generators (WEGs) are usually specified by their *rated power,* which occurs at their *rated wind speed.* At higher wind speeds, power is automatically limited to prevent damage or premature wear and in severe storms most WEGs are shut down completely. WEGs only occasionally achieve their rated power because the wind does not often blow at such high velocities as the rated wind speed. Hence the average power will be typically about 25% (or less) of rated power, depending on the local wind regime and on the rated wind speed of the WEG. It is therefore not easy to compare the performance of different machines unless they have the same rated wind speed.

The gross energy output of a WEG is the most important criterion for the user, but this is a function not just of machine design but of the wind regime at a particular site. The mean wind speed at a site has a profound influence, power and energy being related to the *cube* of the wind velocity. Hence it only requires a 25% increase in wind speed to give almost a 100% increase in energy availability, (and vice versa).

WEGs are available on the international market, or under development, at power ratings from 30 W (0.03 kW) to 5,000 kW, with rotor diameters between these extremes of 0.5 m up to 100 m. The optimum (for converting wind power into shaft power) efficiencies of all reasonably successful wind generators are 30–40%, (larger machines being toward the upper end of this range, and vice-versa).

There are two main classifications for WEGs:

stand-alone systems (for autonomous applications)
grid-connected systems

Most small systems (under 20 kW) are for stand-alone applications and most large systems (over 100 kW) are for grid connection. Medium-sized machines in the 20–100 kW range tend to be mainly used for grid connection, but some are incorporated in stand-alone systems, sometimes in conjunction with diesel-generators as wind-diesel hybrids.

Small wind systems are appropriate for use where they show direct economic or operational advantages compared with other small prime movers. Grid connected systems, on the other hand, are used to supplement fuel-burning generating plant and thereby to save fuel. To this extent they simply represent an investment in capital equipment, which serves to reduce recurrent costs of electricity supply. To some extent they may also avoid the cost of additional thermal plant, particularly when, as in California, the summer demand peak coincides with the peak wind energy availability. In this case the wind systems savings is said to include capacity credit.

STATE-OF-THE-ART

The vast majority of WEGs are conventional horizontal axis machines (i.e., like a fan or an aircraft propeller), but a number of vertical axis turbines have been introduced recently (Darrieus or eggbeater-type and varible geometry-type). The majority of commercially successful medium and large WEGs are two-or three-bladed, horizontal axis machines, with upwind rotors. There is much more variety in the configurations of small machines.

WEGs first came into general use in the 1920s and 1930s mainly on remote farms off the grid in North America and Europe. There has been a major resurgence in the commercial manufacture of WEGs since the late 1970s. Today there are about 100 wind generator manufacturers worldwide, of which around 40 are in the United States. Many are small single-product companies still at an immature stage of development, although several have grown to a significant size very rapidly.

Developing countries have been slow to take advantage of WEG technology. The only developing countries with a WEG industry of any significance are Argentina and the Peoples' Republic of China, although in both cases the industries remain quite small. India has recently initiated a major appraisal of the value of WEGs and is importing a number of Danish machines to build the first Indian wind farms.

There are economies of scale in wind generator manufacture, but there are even larger economies in mass production. Therefore the most cost-effective machines are those that are not so large as to be produced in very small numbers (such as the various megawatt-size prototypes) but are large enough to be cost-effective. These were initially around 20–50 kW, but the size is creeping up to around 100–500 kW at the time of writing.

Similarly, wind power is becoming competitive and convenient for very small scale applications (usually involving battery charging) in areas that are adequately windy. Therefore there are two main "boom" areas in WEG commercialization, namely medium-size grid-connected systems (50–100 kW) and at the other extreme, very small units (50–200 W) for charging batteries.

The four main size categories of WEGs are:

1. Large machines (0.5–5 MW). None of these machines is commercially in series production and in all cases they are intended for use in wind farms to feed the national electricity grids.

2. Medium machines (50–500 kW). Most of this category is applied in wind farms for grid feeding, although some are beginning to be tried either for isolated "mini-grids" or in conjunction with diesel-generating plants or for dedicated stand-alone duties, such as ice production or desalination in remote areas. Although wind farms for grid feeding are outside the scope of this book, the dramatic growth of U.S. wind farms over the last few years (to 1,300 MW by the end of 1987) has had a profound influence on WEG development in this size category, which is of relevance to medium-size, stand-alone and mini-grid WEGs since similar equipment can serve both purposes.

   The efficiency and reliability of these medium-size machines have significantly improved thanks to the hugh increase in operational experience being gained, so that productivity is continuously increasing. Costs of WEGs deployed in wind farms are also decreasing due to the scale of production, for example, in California from $3,100/kW in 1981 to $1,860/kW in 1984. By 1986, WEGs were being installed at costs of around $1,000/kW, and selling power to the grid at $0.07–0.08/kWh.

3. Small systems (1–50 kW). A significant, but much smaller market is for WEGs of 10–50 kW used in particular by many Danish farmers, who can get tax credits that make it economically attractive for them to generate their own electricity and to sell any surpluses to the electricity utility.

   Machines of this size range are appropriate for many isolated integrated wind-diesel applications, where the WEG can save diesel fuel and extend the life of the diesel generator by taking over whenever sufficient wind is available. Systems of this kind are under development, *but they cannot yet be considered to be commercially available.* It is likely to be several years before various control problems are resolved and truly proven and reliable, automatic wind-diesel systems are available.

   There is also a significant but smaller market mainly in the United States, for WEGs of 1 to 2 kW used for battery charging in situations such as remote holiday homes off the main routes. Several U.S. manufacturers of WEGs in this size range claim to have sold several hundred machines and a few have sold over a thousand. Europe,—France, in particular,—has several manufacturers who address a similar market to the United States.

4. Micro WEGS (20–800 W). These machines are direct-driven permanent magnet alternators and are mechanically simple and easy to install. They compete directly with solar photovoltaic systems for such purposes as small-scale lighting, telecommunications, navigation aids, running TV or radio, and such. They are quite widely used on small boats to keep batteries charged. They have the advantage of being more cost-effective than solar PV systems, even in sunny regions, wherever mean wind speeds exceed about 4 m/s and they are also more practical in areas with limited sunshine at certain times of the year (e.g.,

temperate and high latitudes in winter) or where snow or dust can obscure solar-powered systems and render them inoperative.

The performance and reliability of these small machines have significantly improved (they need little or no maintenance) and their costs are stable or decreasing. There is considerable potential for major reductions in cost (perhaps to 30–50% of present levels) once the volume of production increases sufficiently for real mass production.

The current market is in the region of 5,000 or more units per annum of machines of this kind, particularly in the 50W-size range, and a sizeable fraction of these already sells to developing countries, although none are as yet manufactured in developing countries.

CONSTRAINTS TO WIDESPREAD DIFFUSION

The most important constraint is whether there is adequate wind at a specific location. Resolving this all-important question is often complicated by lack of reliable data on actual wind regimes.

Another present constraint is that the market has still not reached the level where true mass production of wind turbines becomes feasible. There is the potential for a significant reduction in costs, which will make wind power more attractive as an option. Ignorance in the market is also a problem. Few potential users will even consider wind systems as manufacturers have so far failed to achieve the levels of promotion and publicity needed to create a more general awareness of their products.

The main technical constraint to the more widespread use of wind energy technology relates to lack of an adequately developed wind-diesel generating system. However, small WEGs for battery charging and larger machines for feeding grids or mini-grids are technically adequate, and so are wind pumps.

A serious constraint to the use of WEGs for feeding grids or mini-grids is that it is essential that the penetration of wind-generated electricity is limited to around 30% even at times when power demand is small. Hence in most cases the WEG capacity needs to be limited to around 10% of total capacity and the potential for fuel saving is therefore quite small, although sometimes worthwhile nevertheless. The reason for this limitation is to maintain voltage stability in the system, but developments in control technology may permit higher penetration levels to be achieved for WEGs in a few years time. Similarly, medium-size dedicated or packaged applicaitons for WEGs, such as ice production or desalination, are under development and will probably become commercially viable within the next decade.

COSTS

The easiest way to link wind turbine costs and performance is through the parameter of a rotor swept area. Surveys of WEG costs have indicated that the cost per unit area of motor follow a consistent pattern as follows:

Rotor Diameter (meters)	Cost ($/m2)
0.5–3	400–800
3–5	350–700
5–12	300–500
12–40	200–600
over 40	400–1000

It can be expected that mass production of the smallest machines will bring their unit area costs down to around $200–500/m2. But even at equal unit area costs, the larger machines have an inherent advantage as they are slightly more efficient and also are mounted higher so as to intercept a higher mean wind speed at any given location. Mechanical wind pumps, on the other hand, typically cost $200–500/m2 in the 2–8-m size range and therefore are already more economic than WEGs in terms of cost per unit of wind energy converted.

Stand-alone systems with battery storage need a sizeable investment in batteries. Typically one week's storage capacity may be needed, but apart from consideration of storage requirements, a major constraint in sizing a battery system is the maximum charging current that often dictates the minimum storage capacity. Typical battery costs at present are in the region of US$ 150–200/kWh of storage capacity. Typically this results in a battery storage costing around 30–70% of the cost of the wind turbine itself. Control system, cabling, power conditioning, and installation when added to battery costs will usually result in a total installed system cost from 50–100% above the ex-works cost of the wind turbine and tower alone.

Wind-diesel hybrid systems also carry a significant cost overhead in short-term storage (usually some batteries needed to smooth the WEG output), plus power conditioning equipment, switchgear, controllers, cabling, and so on. It can be expected that this will amount to an additional cost exceeding 50% of the basic wind turbine. Grid-connected systems have a much smaller installation and balance of system overhead (assuming the grid is nearby). Typically the overheads for a wind farm might realistically be 20% over the wind turbine costs.

Operating and maintenance (O&M) costs for wind generators are generally low for good quality machines and declining as the technology improves. Micro-size WEGs (20–800 W) are nearly maintenance-free and need virtually no O&M (battery maintenance being the main requirement except where maintenance-free batteries are specified). More complex small systems (1–10 kW) are estimated at having O&M costs around 5% of capital cost. Larger machines, especially grid-connected ones, are reported to have O&M costs of around 3% of the capital cost, although stand-alone systems with more complex interfacing with diesels, for example, are likely to have higher percentage O&M costs. Mechanical farm wind pumps typically need one overhaul per year, which may represent an O&M cost of about 1–5%.

The ultimate criterion is the unit cost of electricity generated (or water pumped). Assuming adequate to favorable locations (annual mean wind speed in the 5–7 m/s range), then unit output costs ranging from around $1,00/kW for the smallest

Table 18.1
Wind Generator Unit Output Costs

Rating (kW at 10m/s)	.1	1	2	10	25	50	100
Windspeed at hub-height (m/s)	5	5	5	6	6	6	6
Capital cost ($/m2)	700	600	500	450	400	350	300
Efficiency coefficient	.15	.18	.2	.22	.24	.24	.24
Rotor area (m2)	0.67	5.56	10	45	104	208	417
Rotor diameter (m)	0.92	2.66	4	8	12	16	23
Wind gen. cost ($)	467	3,333	5,000	20,455	41,667	72,917	125,000
Total installed cost ($)	700	5,000	7,500	30,682	58,333	94,792	150,000
Operating life (years)	10	10	10	15	15	15	20
Annualized cost at 10% dis. rate	114	814	1,220	4,034	7,670	12,462	17,625
Average output (kW)	0.01	0.12	0.25	2.16	5.40	10.80	21.60
Output during life (kWh)*	547	5,475	10,950	141,912	354,780	709,560	1,892,160
Cap cost life (c/kWh)	128	91	68	22	16	13	8
Cap cost 10% disc. rate	208	149	111	43	32	26	18
O&M costs ($/yr based on 4% of cap)	28	200	300	1,227	2,333	3,792	6,000
O&M component (c/kWh)	51	37	27	13	10	8	6
Net unit cost (c/kWh)(life cycle)	179	128	96	35	26	21	14
Net unit cost (c/kWh) 10% disc.	259	186	138	56	42	34	24
Average output (kWh/day)	0.15	1.50	3	26	65	130	259

*Assuming average load factor of 0.5.

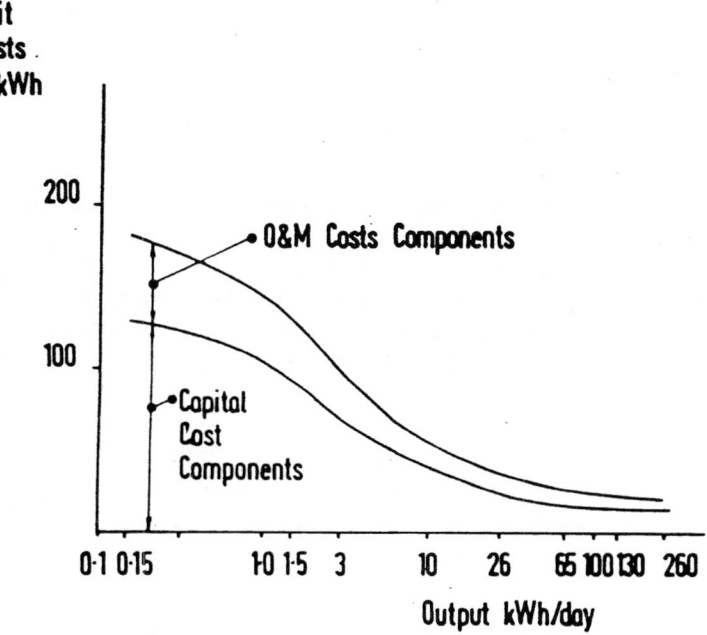

Figure 18.1 Wind generator unit output costs.

micro-wind generators in adequate winds (rather than favorable ones) down to less than $0.10/kWh for large wind farm machines in favorable wind conditions, are feasible (using conventional financial criteria).

Table 18.1 gives an indication of capital and energy delivered costs for wind generators rated from 100 W to 100 kW. Average wind speed at rotor hub-height was assumed at 5–6 m/s. Costs are also shown graphically in Figure 18.1.

WIND/DIESEL ICEMAKER

As an example of a wind/diesel hybrid, the icemaker installed at the fishing village of Abu Ghosun on the Red Sea coast of Egypt in early 1987 with UNDP/UNDTCD funding is of interest. Ice is required both to preserve the quality of the fish for marketing and for prolonging the stay at sea of small fishing boats. The value of the ice in terms of incremental returns on the sale of fish can be very high indeed, remembering that 1 ton of ice per ton of fish may be sufficient to increase the market value by 20% or more.

Requests for proposals (RFPs) were made on the basis of a design study for the required production of 3 tons/day of ice from seawater. By incorporating an oversize icemaker and ice store, the load can follow the energy input (determined by wind). The icemaker has a dc motor (instead of ac) but is otherwise standard. A control system controls icemaker, battery charging, diesel start and stop, and the WEG output. The system draws 29 kW (corresponding to 8.5 m/s wind speed) at full load including 22 kW for the compressor 2 kW for water pumps and auxiliary services and the balance for battery charging (100 kWh of buffer storage). At lower wind conditions the icemaker can operate at half-load (4 cylinders instead of 8 in the compressor) and draws only 16 kW. At high wind speeds, output from the synchronous generator is held constant by increasing the current through the field windings, thus slowing the rotor into an aerodynamically less efficient part of its characteristic curve.

A diagram of the system is shown in Figure 18.2 and a cost breakdown in Table 18.2. In Table 18.3 the cost of ice is shown based on different assumptions. Because the wind regime is not particularly good at the site, the wind/diesel system could only compete with a conventional diesel icemaker given a low (5%) discount rate rather than 10% more often used by the World Bank, for example, in this kind of economic analysis).

The particular wind machine chosen (Wincon, ex-Micon 55) has an outstanding record in California wind farms where it routinely exceeds 30% annual capacity factor and a high reliability with an average of 4.2 unscheduled outages annually for a total of 70 hours (including scheduled maintenance) or better than 99% availability.

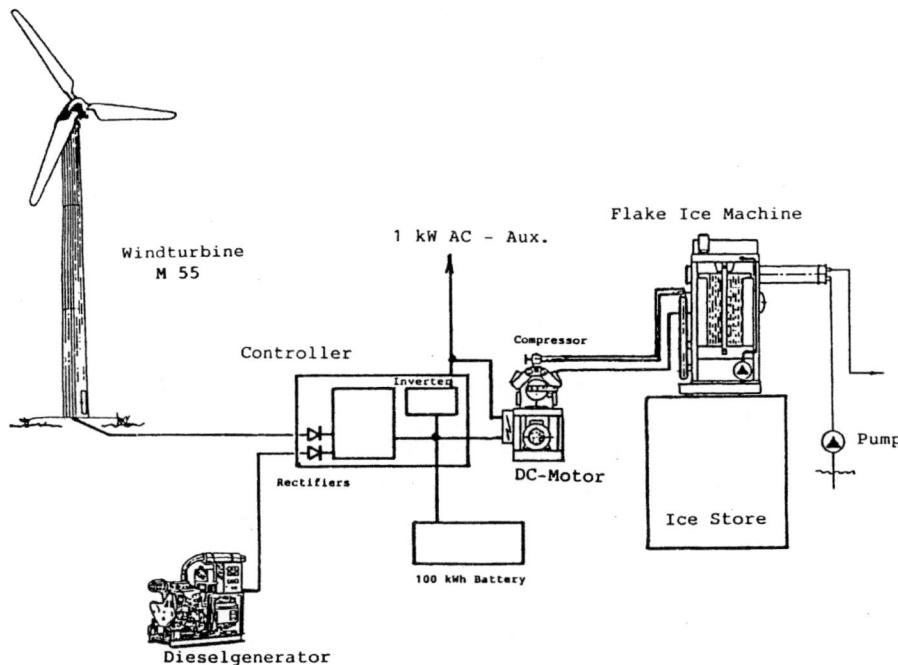

Figure 18.2 Wind icemaker schematic.

Table 18.2

Wind Icemaker Cost Breakdown: UNDP/UNDTCD Project in Egypt Site. Abu Ghosun (Red Sea) (Annual average wind speed at hub-height 5.5m/s)

Item	Cost US$*	%
Wind generator 55 kW at 14m/s	$ 47,500	23
Seawater icemaker 8.5 ton/day		
(300 Vdc 22 kW 68 wKh/ton)	56,700	28
14-ton ice storage	5,200	2
Diesel generator 32 kW	9,200	4
Battery 100 kWh	18,200	9
Control system, instrumentation,		
misc. hardware	32,100	16
Installation, training, warranty	24,500	12
Local costs (civil works)	5,000	2
CIF	5,600	3
Total	$204,000	100

Annualized cost at 10% and 5% discount rates	10%	5%
7 1/2-year depreciation for batteries	$ 12,200	9,700
20-year depreciation for balance including icemaker, wind generator, and diesel (running 1,000 hr/yr)	15,200	10,400
Diesel fuel cost based on $0.14/kWh and 68 kWh/ton ice and 219 ton ice annually from diesel power	2,100	2,100
O&M at rate 2% capital cost	4,100	4,100
Total annual costs	33,600	26,300

Ice production:
3 ton/day × 365 day × 0.80 (wind fraction) = 876 ton by wind
+ 3 ton/day × 365 day × 0.20 (diesel fraction) = 219 ton by diesel
cost of ice $33,600/1095 ton = $31/ton   (10% disc. rate)
cost of ice $26,300/1095 ton + $24 ton   (5% disc. rate)

*Prices converted from Dkr at 9 Dkr = 1 US$.

Table 18.3

Comparative Cost of Ice at Abu Ghosun Under Different Assumptions

Production:	3 ton/day from seawater
Wind:	5.5 m/s annual average speed. Average output from wind machine 170 kWh/day corresponding to 15% capacity factor. Wind fraction 0.80 diesel fraction 0.20
Diesel fuel:	$0.40/liter

Equipment costs, O&M, depreciation, and discount rates as Table 18.2.

1. Wind/diesel icemaker	$31/ton
2. As above discount rate 5% (not 10%)	$24/ton
3. Conventional diesel icemaker	$22/ton
4. Diesel icemaker fuel $0.80/1 (not $0.40)	$32/ton
5. Diesel icemaker fuel $0.20/1	$17/ton
6. Wind/diesel icemaker, 7–8 m/s av wind (30% c.f.)	$15/ton
7. Price of ice delivered by road from Cairo*	$28/ton
8. Commercial price of ice in Cairo*	$14/ton

*1 Egyptian pound = US$0.70.

POTENTIAL FOR LARGE-SCALE DEVELOPMENT

WEGs are today generally seen as one of the most promising alternative means for generating electricity, and hence are receiving major R&D support in the energy programs of many of the industrial countries. The technology probably has a good future in developing countries because many of them have apparently favorable wind regimes and high conventional generating costs. Also, the technology even in its present form demands few skills or manufacturing facilities that are not already in place in the industrial sectors of many of the middle level and more advanced developing countries; therefore, WEGs have the potential for significant import substitution both in terms of fuel and in terms of generating plant.

ADVANTAGES AND DISADVANTAGES

The main advantages of WEGs are:

1. Economic and adequately windy areas.
2. Widely applicable in some regions.
3. High degree of autonomy can be achieved with small units (only limited human intervention needed compared with small engines).
4. WEGs have a long operational life compared with small engines.
5. Local manufacture feasible in developing countries.

The main disadvantages are:

1. Totally uneconomic in poor wind regimes; applicability limited to windy regions only.
2. Requires open, unobstructed terrain, free of trees, or sharp-edged topography.
3. Output fluctuates greatly (so power conditioning is often needed between WEG and the load).
4. Longer term fluctuations demand energy storage or an application where alternative energy sources provide backup power during calms.
5. Only limited penetration of grids or of mini-grids is feasible at present.

CONCLUSIONS

Wind generation of electricity deserves serious consideration in areas with adequate wind regimes (typically having mean wind speeds exceeding 3 m/s for farm wind pumps and 4–5 m/s for WEGS).

It is important that data collection and analysis of the wind energy resource is initiated wherever it appears to offer some prospect for practical applications, as the limited state of knowledge of wind regimes is often a constraint.

Stand-alone applications of WEGs or wind pumps are likely to be of interest for rural development, telecommunications, health, agriculture, and private sector users, whereas grid-feeding concerns energy departments and electricity utilities. Coordination could be through the research and development or planning wing of an energy ministry (this is the common approach in industrial countries with wind programs).

Where a good case for the reasonably widespread use of wind system can be made, pilot-demonstration projects are needed initially to confirm the practicability and economy of using wind power in a specific region. Such projects should be planned from the outset to have a budget for adequate performance monitoring and analysis, or little of value will be learned. All too often the monitoring of pilot projects is inadequately funded and therefore fails to exploit the project effectively.

19

Decentralized Electricity Generation and Biomass

ELECTRICITY GENERATION FROM BIOMASS

Among the nonfossil energy sources used to generate electricity, biomass is one of the oldest and most flexible. It is certainly an early application, especially with the production of electricity by agricultural processing industries using their own wastes, including oil mills, sugar factories, and such. These industries thus operate their plants autonomously and also sometimes supply electricity to the surrounding population. It is theoretically flexible, because through the different conversion techniques, it serves to cope with the different power ranges, and to respond in principle to any electrical demand, provided the material is available. Among the available techniques, the two main ones are:

1. Combustion, which involves preparation, combustion for steam generation, steam turbine, ac generator.
2. Gasification, which involves preparation, gasification, gas cleanup, internal combustion engine, generator.

Combustion is a fully proven technique at the commercial level for wood and a number of highly ligneous wastes (groundnut shells, sugar cane bagasse), a technique currently tending to spread to other materials (rice husks, cottonseed, pods, etc.) (see Table 19.7). Its main drawback is the scale factor: if capacity decreases, the capital cost per kW installed capacity rises sharply, whereas the efficiency of the installations drops.

Below 1 MW, combustion is no longer competitive in principle with other alternatives such as gasification and is only economically justified in case of cogeneration (demand for low pressure steam for industrial use, for example). The efficiency of gasification in these power ranges is far superior to that of direct combustion (20–25% compared with 10–15%). In power ranges of several hundred kW and above, however, gasification is still in the developmental stage. A few installations exist, but the commercial stage lies in the distant future.

Small gasifiers in capacities of a few dozen kW, which are often derivatives of the simple techniques used in World War II, can, however, be considered as having reached the commercial stage on materials such as charcoal, wood, and coconut shells. They are still experimental on other materials (apart from rice husks, on which Chinese gasifiers are routinely operated, but in a specific environment). A number of other more marginal systems also exist:

1. Pyrolysis (anerobic gasification), whose value is mainly limited to the production of secondary fuels—coals, tars—in addition to gas.
2. Liquid biofuels such as alcohols (ethanol, methanol), and vegetable oils and their esters, whose production and combustion in a motor is technically proven but whose use for the production of energy is generally faced with cost constraints in comparison with conventional fuels. The large-scale production of ethanol, as a motorfuel, as a national policy (as in Brazil) is not considered here.
3. Biogas, produced by the methane fermentation of organic matter (animal wastes, wet agricultural and industrial wastes). This can be used by direct combustion, for lighting, cooking, or in a motor. Its use to generate electricity is faced with a number of handicaps (relatively low energy efficiency, large tank volumes) and is only justified in very specific cases, such as in conjunction with environmental cleanup.

CONSTRAINTS AND COSTS

Two configurations can be distinguished:

1. The "support" existence of an agricultural industry, with a large quantity of precollected agricultural wastes, capable of using these wastes for its own electrical needs and of distributing the surplus.
2. The establishment of a stand-alone electricity generating unit, securing its own supply of agricultural material.

The value of a system relying on the existence of an *agricultural industry* derives from:

1. The existence of a large quantity of precollected wastes, often at negative cost—they must be removed—in superabundance (a rice mill theoretically produces five times more energy in the form of wastes than it consumes, a cotton mill seven times more, etc.).
2. The economic value for the industry itself of employing techniques to produce energy from its wastes: the payback time of similar operations in Europe is

about 3–4 years, and one operation in Mali (Huicoma oil mill) has achieved a payback time of about 1.5 years.

The main constraints result from:

1. The large capital investment required (up to $1 million for an additional investment in a waste system), which may require public backing.
2. The negotiation of agreements between the agricultural industry and the national electrical utility for the distribution of electricity in accordance with public service requirements.

For stand-alone power stations, the problematics are quite different depending on equipment size:

1. For high and medium capacities (wood-fired plants of several MW and even a few hundred kW), the supply of wood amounting to thousands and even tens of thousands of m3 constitutes the major constraint.

   In most cases, it appears necessary to install resource management schemes incorporating the condition for its restoration (plantations). The Philippine dendrothermal program (3 MW power stations supplied by cooperative plantations of 11,000 hectares) is a rich source of information in this respect and shows: a theoretical profitability in comparison with diesel thermal plants (between 60 and 80% of the diesel kWh price), and the resource accounts for 40% of the kWh price.

   In fact, the many problems encountered in plantation programs may imply a wood cost that is closer to twice the theoretical cost, accounting for nearly two-thirds of the kWh production cost. Without totally jeopardizing the profitability of this type of project, it nevertheless limits it.
2. For low capacities (a few tens of kW), the problem is quite different and lies in the organization and management capacity of the beneficiary communities.

   The quantities of material involved are limited in this case (no more than the equivalent of wood supply for cooking meals for a few families) and largely available in suitably wooded surroundings.

   Several project and contractor documents reveal the profitability of small wood-fired power stations (gasifiers) in remote areas, compared with the competition from diesel or from connection to grid (as soon as the distance exceeds 10 km), and payback times, which vary between 1.5 and 3 years, despite an extra investment of 60 to 80%. The analysis of a few operating results, however, encourages greater prudence. At best, one can state that the diesel and wood kWh costs are of the same order of magnitude and that the major interest of this type of project resides essentially in the transfer of the cost of imported diesel to wood produced and processed locally. This means a substantial income for the village, often ranging up to several thousands of dollars.

   The essential problem of this type of project is that it cannot by itself support the cost of normal handling by the local electrical utility. At best, the utility will perform periodic maintenance, as part of the overall project, and routine management must be conducted by the beneficiary community. This implies—and this is not always the case—an organization and management capacity that is capable of handling both wood supply as well as routine operation of the power station.

CASE STUDIES

10kW irrigation waterpumping unit. A comparison is made between a 10-kW diesel and charcoal gasifier unit, used for lift irrigation in an isolated area in Ethiopia. Standard characteristics of both units are presented in Table 19.1. A comparative financial analysis is presented in Table 19.2.

Under the locally prevailing circumstances, charcoal gasifier powered lift irrigation has considerable financial advantages over diesel power. Successful introduction of the gasification technology will depend to a large extent on institutional and social factors such as:

Table 19.1

Standard Characteristics of Alternative Waterpumping Units for an Application in Ethiopia

Characteristic	Diesel system	Charcoal gasifier system
Installed capacity (kW)	10	10
Lifetime (yr)	8	8
Initial investment (US$)	5,125	11,900
off-factory	(3,750)	(9,000)
freight, insurance		
installation		
site preparation		
spare parts	(1,375)	(2,900)
Power production (kWh/yr)	27,000	27,000
operating time (hrs/yr)	(3,000)	(3,000)
load factor (%)	(90)	(90)
Fuel consumption		
diesel (1/yr)	10,260	
diesel cost (US$/1)	0.60	
charcoal (kg/yr)		21,600
charcoal cost (US$/kg)		0.03
Labor		
labor cost (US$/hr)	0.50	0.50
labor time (hrs)	270	540
Maintenance and repair		
(US$/kWh)	0.015	0.025
Lubricants		
cost (US$/1)	1.00	1.00
consumption (1/kWh)	0.004	0.006

Table 19.2
Comparative Financial Analysis of Alternative 10 kW Lift Irrigation System (US$/yr)

	Diesel system	Charcoal gasifier system
Annualized capital cost (interest: 10% yr)	961	2,732
Annual operating cost		
fuel	6,156	648
labor	135	270
maintenance and repair	405	810
lubricants	108	162
Total	7,765	4,622
Unit energy cost (US$/kWh)	0.29	0.17

provision of adequate training facilities both on operator as on workshop mechanic level

motivation of (especially) operating personnel in order to overcome initial technical and organizational difficulties and put up with additional (and sometimes dirty) work.

250 kW sawmill power plant. For a medium-size sawmill at a fairly isolated site in Cote d'Ivoire, a comparison is made between the existing diesel power plant, an alternative up-draft gasification system, and a steam system (condensation turbine). Wood residues in a form suitable for the gasifier and the direct combustion system are generated on-site. Both the biomass systems are equipped with adequate storage capacity and automatic fuel handling and feeding equipment. Tarry condensates from the gasification plant are treated in a condensate treatment plant. The steam system is equipped with a water treatment plant of adequate capacity. As no alternative utilization possibilities exist, the cost of wood residue fuel on-site is taken at nil. Characteristics of the different systems are detailed in Table 19.3. A comparative financial analysis is presented in Table 19.4.

Under the prevailing local circumstances, both the up-draft gasifier as well as the steam plant have slightly lower unit energy cost then the existing diesel power plant. Annual profits in case a gasification system is installed amount to US$7,500. For a steam system annual profits are US$6,600. In view of the existing investment, both amounts appear insufficient for a positive biomass-fueled power plant investment decision. Both the gasification system and the steam system would merit very serious consideration in case a replacement investment in power plant was necessary.

The installed power capacity under consideration (about 250 kW) appears to represent roughly the break-even point between gasification and steam systems. At considerable lower capacities, competition is between gasifier and diesel-powered systems. At considerable higher capacities, as a rule, only diesel and steam plant must be evaluated.

Table 19.3
Standard Characteristics of Existing and Alternative Energy Plant at Ivory Coast Sawmill

Parameter	Diesel plant	Steam plant	Gasifier plant
Electricity production (GWh/yr)			
total*	0.55	0.60	0.60
effective	0.55	0.55	0.55
Installed capacity (kW)	250	275	275
Initial investment (US$)	128,500	676,000	532,000
Fuel consumption			
diesel oil (1/yr)	176,000	—	—
wood residue (t/yr) dry basis	—	2,250	750
Labor requirement			
skilled (man year)	1	1	1
semiskilled (man year)	2	2	2
unskilled (man year)	—	4	4
Lubricants (1/kWh)	0,003	0,0065	0,0065
Chemicals (US$/yr)	—	2,500	—
Equipment lifetime (yr)	8	12	8
Maintenance and repair (US$/kWh)	0,015	0,010	0,020

*Both for the steam system and for the gasification system, an internal energy consumption of 0.05 GWh/yr has been taken into account.

1,300 kW cogenerating energy system for integrated sawmill/plywood factory in the Ivory Coast. At present the plant is powered by diesel generators. Process steam for the plywood and veneer factory is generated by means of wood residue fired boilers. As an alternative a steam power system incorporating wood storage, wood hogger, automatic wood fuel handling, wood-fired boiler, backpressure turbine, process steam system, and condensation turbine is proposed. The system is designed in such a way that power and steam supply are guaranteed under all circumstances. As no alternative use for the wood residue exists, wood residue costs are taken at zero. Characteristics of the existing and proposed system are detailed in Table 19.5. A comparative analysis of both systems is presented in Table 19.6.

SMALL EXTERNALLY HEATED ENGINES

A number of recent developments hold promise for small biomass-fueled engine generators.

Air (Stirling cycle) engines. Popular in North America in the last century, small hot air engines (100 W to 10 kW) are now being "reinvented" using modern materials and capable of good efficiency with a variety of biomass fuels including "difficult" ones like rice husks. As an example a 3-kW Stirling water pump is available in India for about Rs 43,000 (US$3,000) or Rs 53,000 (US$4,000) in the engine

Table 19.4

Comparative Financial Analysis of Different Energy Supply Plants for Ivory Coast Sawmill (US$/yr)

	Diesel plant			Steam plant			Gasifier plant		
	local cost	foreign cost	total cost	local cost	foreign cost	total cost	local cost	foreign cost	total cost
Annualized capital cost	4,920	13,680	18,600	9,830	67,070	76,900	20,100	52,400	72,500
Annual operation and maintenance cost									
lubricants	330	1,320	1,650	780	3,120	3,900	780	3,120	3,900
chemicals	—	—	—	500	2,000	2,500	—	—	—
labor	16,200	—	16,200	22,200	—	22,200	22,200	—	22,200
maintenance and repair	4,125	4,125	8,250	3,000	3,000	6,000	6,000	6,000	12,000
Annual diesel fuel cost	25,520	47,880	73,400						
TOTAL	51,095	67,005	118,100	33,310	75,190	111,500	49,080	61,520	110,600
Cost per effective unit of electricity (US$/kWh)			0.21			0.20			0.20

Table 19.5
Standard Characteristics of Existing and Proposed Energy Plant at Ivory Coast
Semiintegrated Wood Processing Facility

Parameter	Existing system	Proposed system
Energy production		
electric power (GWh/yr)	5.6	6.7
process steam (t/yr)	99,500	99,500
Installed capacity		
electric power (kW)	1,100	1,300
steam (t/hr)	12	12
Initial investment (US$)	385,000*	1,972,500
(installed)	650,000[†]	
Fuel consumption		
diesel (DDO) oil (1/yr)	1,820,000	—
wood residue (t/yr)	31,600	48,510
Operating manpower		
skilled (man year)	1	1
semiskilled (man year)	4	4
unskilled (man year)	10	18
Lubricants (1/kWh)	0.003	0,0065
Chemicals (US$/yr)	7,000	11,250
Equipment lifetime (yr)	12*	15
	15[†]	
Maintenance and repair	6*	4
(% of initial investment)	3[†]	

*: Diesel power plant.
[†]: Separate process steam boiler.

generator version and with an efficiency of up to 16%. These engines are now undergoing exhaustive testing and are expected to cover the important range up to 1–2 kW not now adequately covered by alternatives.

Steam piston (rankine cycle) engines. Small steam piston engines (1–10 kW) are also being redeveloped using modern techniques to generate power in the low range with biomass fuels.

Organic rankine cycle engines. Organic rankine cycle (ORC) expanders and turbines using low boiling organic liquids (butane- or freon-type fluorinated hydrocarbons) instead of steam as working fluid, were first developed for use with low temperature heat sources available from solar ponds and solar thermal collectors where their efficiences are necessarily low (under 5%). They have been adapted for slightly higher temperature sources such as geothermal fluids and biomass fuels. One Israeli manufacturer now offers a line of ORC turbine generators, which are highly reliable but also too costly for other than specialized applications (the smallest 1 kW unit cost more than US$15,000).

Table 19.6

Comparative Financial Analysis of Existing and Proposed Energy Plant at Ivory Coast Semiintegrated Wood Processing Facility (US$/yr).

	Existing system			Proposed system*		
	local cost	foreign cost	total cost	local cost	foreign cost	total cost
Annualized capital cost	7,000	49,500	56,500[†]	31,950	227,430	259,380
	9,000	76,470	85,470[‡]			
Annual operation and maintenance cost						
lubricants	3,360	13,440	16,800	8,710	34,840	43,550
chemicals	1,750	5,250	7,000	2,810	8,440	11,250
labor	35,400		35,400	47,400		47,400
maintenance and repair	16,500	16,500	33,000[†]	39,450	39,450	78,900
	9,750	9,750	19,500[‡]	—	—	—
Annual diesel fuel cost	263,900	495,040	758,940			
Total	346,660	665,950	1,012,610	130,320	310,160	440,480
Cost per effective unit of electricity (US$/kWh)			0.18			0.08

*Pay-back period of total initial investment of proposed system: 3.4 years.
[†]: Diesel power plant.
[‡]: Separate process steam boiler.

Table 19.7

Heating Values of Wood, Gas, and Various Agricultural Wastes Compared to Gas-Oil.

	Heating value (LHV)*	Energy use	
		Direct combustion	Gasification
Wood (homogeneous)	14–17 MG/kg	Commercial	Commercial (x10 kW), devt. above
Sugar cane bagasse	16 MG/kg	Commercial	Development
Groundnut shells	17 MJ/kg	Commercial	Development
Coconut shells	17 MJ/kg	Commercial	Commercial (×10 kW), devlop. above
Cotton seed pods	16 MJ/kg	Development	Development
Rice husks	12 MJ/kg	Development	Commercial† (×10 kW), devt. above
Diesel	42 MJ/kg		
Producer gas	4 MJ/m3		
Methane fermentation	23 MJ/m3		
Natural gas	36 MJ/m3		
Butane gas	120 MJ/m3		

*Biomass heating values obviously vary considerably depending on moisture content. The figures are given for a moisture content of about 10 to 15%.
†Chinese models.

CONCLUSIONS

Two alternative actions appear to emerge in the short term:

1. Expansion of the use (in boilers, and with further development, in gasifiers) of agricultural wastes by the agricultural industries, by incorporating the generation of its electricity in the national electrification development schemes. This includes: identification of agricultural industries and volumes of wastes available, and negotiations between the electrical utility and the industrial undertaking concerning investment decisions, financial arrangements, specifications, tariff procedures, etc.

2. If conditions are favorable, the development of programs for wood-fired gasifier micro-power stations of a few tens of kW, in relatively remote areas. This includes: site identification (for programs with a minimum size of about 10 units) in a given region, and the setting up of agreements between the electrical utility and the local communities for routine operation (maintenance, etc.).

20

Renewable Energies and Water Pumping

INTRODUCTION

In rural (unelectrified) areas of developing countries, the simplest means of obtaining drinking water is to draw it directly from a river or open well using a rope and bucket and then carrying it to the point of consumption. A large proportion of villagers' (usually women) time is often spent collecting water. Traditionally, improving the water supply is accomplished by drilling a borehole and fitting it with a manual pump. This provides a cleaner and more reliable supply, but a relatively large amount of labor must still be expended pumping and carrying the water. Where larger quantities of water and/or a piped distribution is required, it is normal to use a diesel engine driven pump.

In some areas of the world wind pumps have been used for many years, whereas solar pumps have recently emerged as a viable technology. There is great scope for widespread use of these water pumping technologies. They can provide better service including water distribution than manual pumps, and they replace diesel pumps, giving higher reliability. Wind and solar pumps are more expensive to buy than conventional pumping systems, but under certain conditions they have lower overall (life-cycle) costs than diesel pumps.

All pumping systems can be characterized by the flow of water (liters per second, cubic meters per day) and the head through which the water is pumped (meters). Figure 20.1 indicates the range of water pumping duties that can typically be met using solar, wind, and human- and animal-powered pumps.

Water pumping for irrigation is not dealt with in this book. The value of irrigation water is very much lower, based on the value of crops grown, than drinking water (save in exceptional circumstances, e.g., where small quantities of high value vegetables are grown). Hence pumping by any means that has to be paid for is unlikely to be economic. Wind pumps and solar pumps are used almost exclusively for water for human and/or animal consumption. However, the benefits from using "excess" water from a wind or solar pump for irrigation of small gardens can be significant and the use of excess water must be foreseen and studied in water pumping projects.

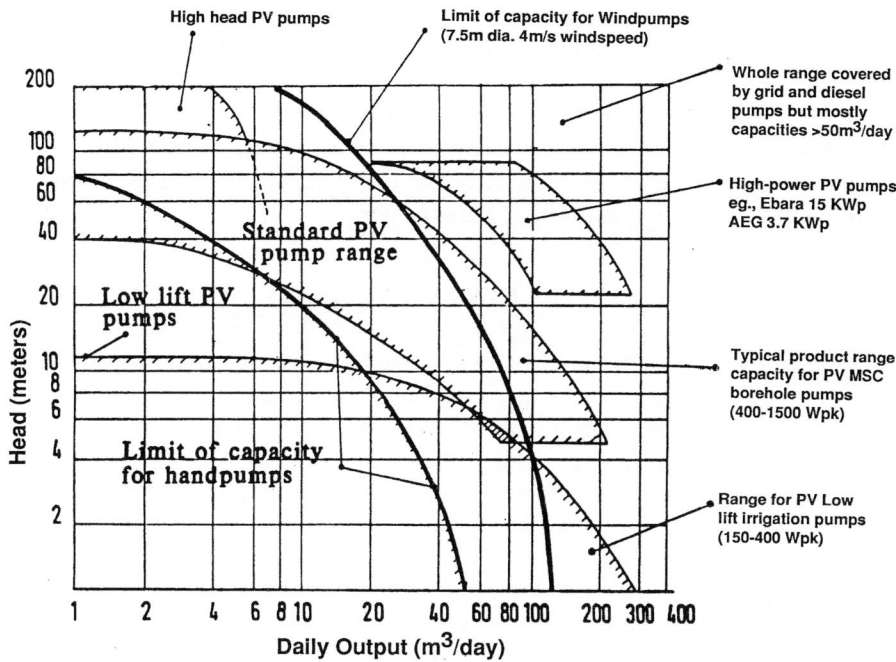

Figure 20.1 Pumping duties for various systems.

Surplus water enables self-development of the village with economic activity induced by gardens and the sale of water.

A target for water consumption is 40 liters per person per day. A typical village of 500 people needs 20 m3 per day. Pumping heads range from 10 m to 60 m. The output from solar and wind pumps changes with the season. It is necessary to ensure that the water requirement can be met in the least sunny or least windy month.

STATE-OF-THE-ART OF SOLAR PUMPS

There are at present at least 2,000 photovoltaic pumping systems installed around the world. This is a large number by renewable energy standards but minute compared with the potential. Many are known to be working to the satisfaction of their users, although unfortunately few of these have been monitored. Several manufacturers have undertaken the research and development required to produce products that are highly efficient and sufficiently robust and reliable for use in the field.

There are a variety of different designs. In many cases it is feasible to utilize off-the-shelf, mass-produced motors and pumps. However, special pumps and motors have been developed by some manufacturers with an above average efficiency to minimize overall system costs. The efficiency of conversion of electrical energy from the array to delivered hydraulic energy is between 30 and 40%. (3–4% solar to hydraulic energy if we include 10% solar to electric array efficiency.) Photovoltaic pumps are rated in peak watts—that is, the power output under reference peak sunlight and with a cell temperature of 25°C.

Modules are available in sizes ranging from 2 to 50 watts. Large power outputs from a single source can be obtained by combining modules. The size of a system for a given use depends on the available solar radiation. Smaller/bigger villages will have proportionately smaller/higher size systems.

STATE-OF-THE-ART OF WIND PUMPS

Wind pumps are windmills specifically dedicated to pumping duties. Although wind electricity may be used to power an electric pump, this is unusual and most are purely mechanical with a rotor driving a piston pump. Around 1 million farm wind pumps are currently in use worldwide. Although they tend to be heavier for their size than wind generators, their cruder construction results in lower costs per square meter of rotor area, and hence they tend to be economic in mean wind speeds as low as 3 m/s compared with alternatives.

Wind pumps are commercially available in sizes from about 1.5 m up to 8 m diameter from some 50 manufacturers worldwide. Several developing countries have significant wind pump manufacturing operations, including Argentina, Brazil, Colombia, Kenya, Zimbabwe, Senegal, Pakistan, India, China, Thailand, and the Philippines. Australia and the United States are the main industrialized countries producing wind pumps.

Many of the commercial products were developed early this century and have changed little in design for decades. They are still heavy and complicated to install, but highly reliable and robust if provided with limited maintenance. Various lighter, more modern designs have recently appeared and are beginning to be marketed. Maintenance required is regular lubrication of shaft bearings (from one to four times annually) and change of pump leathers (every 2 years).

The power provided by a wind pump is proportional to the swept area of the rotor. The wind pump size for a given duty depends on the wind speed for the location. Figure 20.2 shows typical system sizes for a village of 500 people at three different wind speeds for different pumping heads.

Average hydraulic power, P in watts (flow × head) of a wind pump is given quite accurately by:

$$P = 0.1 \, AVP + 0.1 \, AV.$$

A is swept area in square meters
V is the cube of average wind speed in m/s

This assumes that the pump has been correctly sized for average wind speed and head. For a given head and pump, the hydraulic power (discharge) is directly pro-

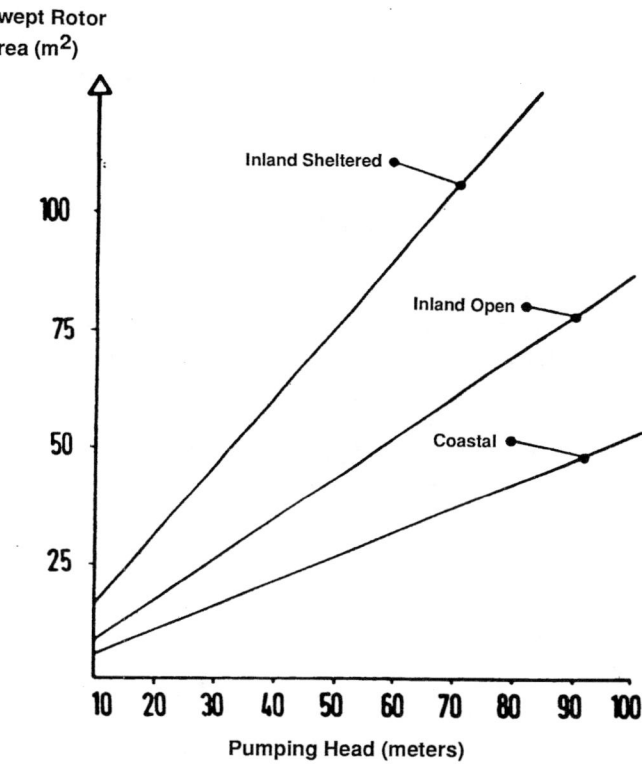

Figure 20.2 Wind pump sizes (in terms of swept rotor area) for a village of 500 people (20 m³/day).

portional to average wind speed. Different village populations will require wind pumps sized in proportion to the swept rotor areas (see Fig. 20.3). For example, a village of 250 people will require half the rotor area shown in Figure 20.3.

CONSTRAINTS TO WIDESPREAD DIFFUSION

The principal constraints to the widespread use of wind and solar pumps are:

Financing of relatively high capital costs.
Lack of local infrastructures for distribution of spare parts and maintenance.
Lack of information available on the true costs of using alternatives such as diesel pumps.
Credibility of "new" technologies with users.
Lack of data on storage requirements.

Wind and solar pumps have received a lot of attention in many developing countries, perhaps raising necessary awareness but also raising expectations. Unfortunately, in most countries few decision makers have sufficient experience with them, especially concerning their economic and financial readiness to meet some specific needs in their own countries. Until this lack of specific information is remedied, it will be very difficult to finance larger scale projects and it will be equally difficult, indeed impossible, to facilitate sales financed by a country's internal resources.

COSTS

For *capital costs,* solar pumps at present range from $9 to $15 per watt. Smaller systems are more expensive per watt of installed power; for example, a 500-watt system may cost $12 per watt, whereas a 1,500-watt system may cost $9 per watt.

Wind pump costs vary significantly from country to country, from $150 per square meter of rotor area in India to $1,400 per square meter of rotor area in Germany. More typical costs are $200 to $300 per square meter of rotor area.

Capital costs for a given water requirement will depend on the local climate. Typical values for a village of 500 people are shown in Figure 20.3.

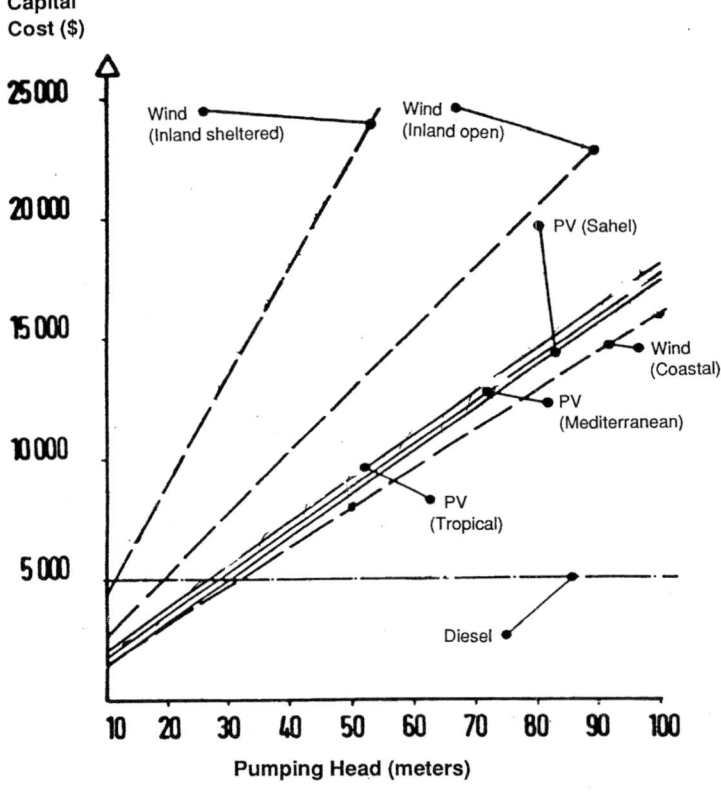

Figure 20.3 Capital costs of solar, wind, and diesel pumps for a range of pumping heads and different climatic conditions (20 m³/day requirement).

Unit costs (i.e., $ per cubic meter of water) are obtained by carrying out a life-cycle costing. Future costs are discounted to the present. The discount rate reflects the opportunity cost of capital.

The unit cost is based on the water required and not the water that can be pumped (since a solar or wind pump has the potential to provide excess water in periods of high sun or wind). Also, in most irrigation applications water is required for only 90–120 days annually, which puts both wind and solar pumps at a disadvantage compared to diesel, where running costs are more significant than capital costs.

Lifetime operating and maintenance costs are required for a life-cycle costing. Typical values for solar, wind, and diesel pumps are shown below. For solar versus diesel pumps, a number of studies suggest a break-even of around 1,000 m^4/day assuming year round demand, and a break-even for wind versus diesel pumps of 2,000–3,000 m^4/day for winds of 3 m/s or better.

Unit costs for solar, wind, and diesel pumps are shown in Figure 20.4. This shows that under good conditions solar and wind pumps deliver water more cheaply than diesel pumps. Comparisons with manual and animal pumps are much more difficult.

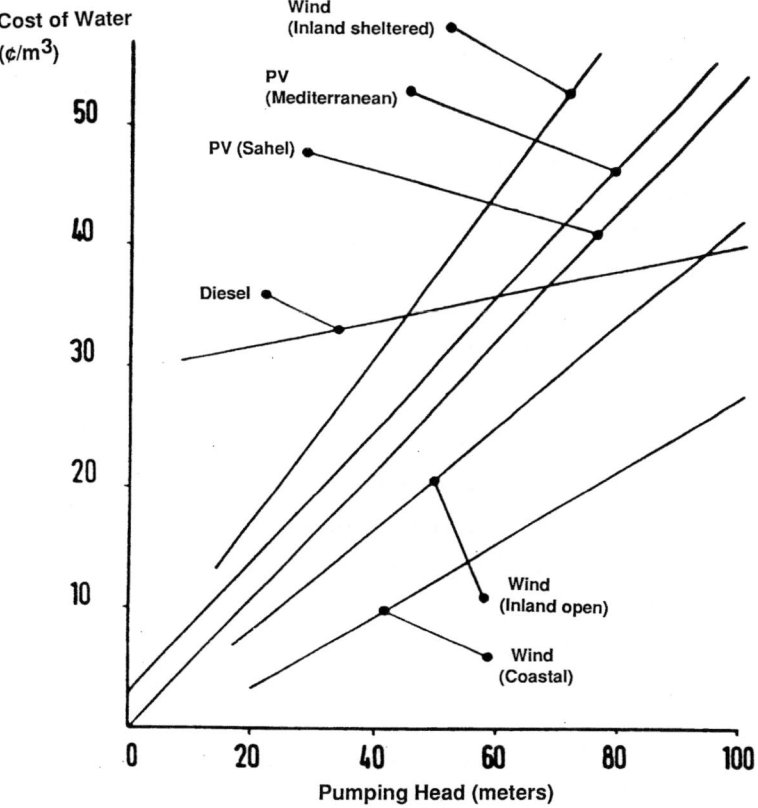

Figure 20.4 Unit costs of solar, wind, and diesel pumps for a range of pumping heads and different climatic conditions (20 m^3/day requirement).

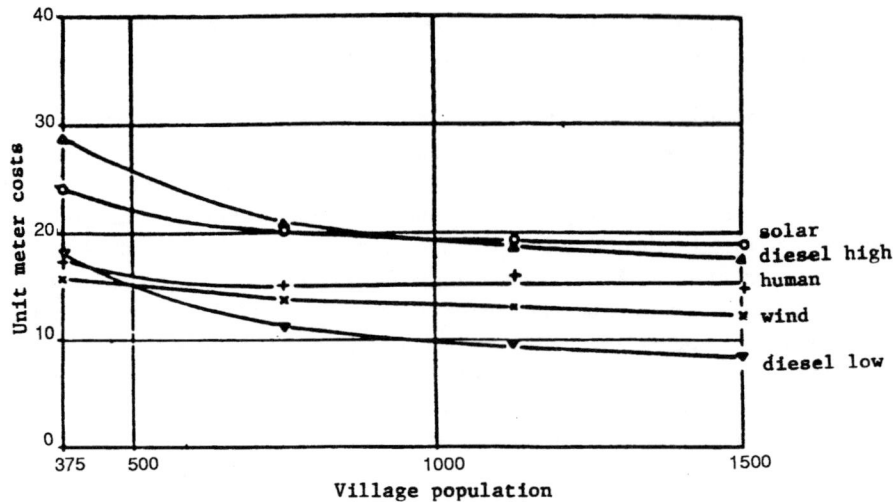

Figure 20.5 Cost of water for village supplies—no distribution (Baseline scenario—20 m static lift; UNDP/World Bank).

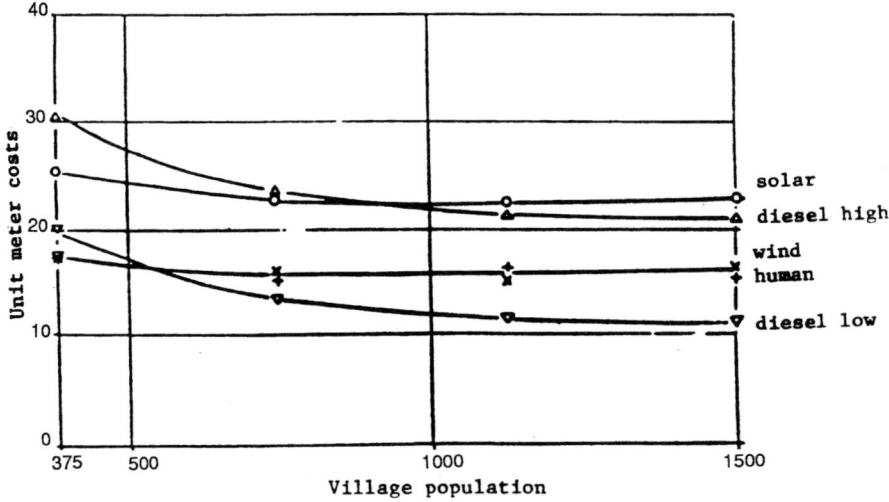

Figure 20.6 Cost of water for village water supplies—piped distribution (Baseline scenario—20 m static lift; UNDP/World Bank).

Figures 20.5 and 20.6 show some results from the UNDP/World Bank Global Solar Pumping Project.

ADVANTAGES AND DISADVANTAGES

The main advantages of solar pumps are:

   1. Minimal maintenance, unattended operation.

2. No fuel required.
3. Small operating and recurrent costs.
4. Easy installation.
5. Predictable performance.
6. Long life.
7. Often economic compared to diesel.
8. High modularity (from 1 to 1,000 m3/day and 1 to 100 m head).
9. Can be part of larger PV systems.
10. Use of excess electricity possible (e.g., for lights).

The main advantages of wind pumps are:

1. Long life.
2. Maintenance relatively simple (compared to diesel).
3. Economic in windy areas.
4. Local manufacture is feasible.
5. Long experience.

The main disadvantages of solar pumps are:

1. High capital cost.
2. Limited experience.
3. Periods of low output.
4. Equipment selection is at present not straightforward.

The main disadvantages of windpumps are:

1. Output is highly sensitive to wind speed.
2. There may be (windless) periods with zero output.
3. Require open unobstructed terrain.
4. Installation and transport cost higher than solar pump.
5. Wind data often not available.

21

Renewable Energies and Domestic Energy Needs

STATE-OF-THE-ART

Low capacity (less than 500 W) photovoltaic generators for domestic use are distinguished by great simplicity. They typically comprise the PV modules with their support frame the charge-discharge regulator, and batteries. The dc electric power supplied is usually at 14 or 24 volts, and is used to run lamps, a refrigerator, television set, and so on. Small PV generators for domestic use are mature, proven products today, giving overall satisfaction if the size and choice of components and the training of the users are properly carried out.

Substantial improvements can be anticipated in the medium-term, reinforcing reliabilty, and above all reducing costs: These improvements include new technologies for modules, improved low-maintenance batteries, more reliable regulators, and better electrical appliances. The scale effects may be important in this case: group purchases to achieve a significant cut in module price, standardization, reduction of overheads and commercial costs, installation by the user thanks to the supply of kits, and such. This approach has been successfully employed in the Polynesia solar program (case study).

It is possible to use locally made car batteries with regular replacement of lead plates instead of long-life deep discharge PV batteries conventionally used. The low cost and saving of foreign exchange may compensate for the poor characteristics of the batteries. At the other extreme, nickel-cadmium batteries suitable for PV applications, with 20-year life and capable of 100% discharge with virtually no maintenance, have recently appeared and may prove cost-effective in some applications despite their high cost ($500/kWh versus about $200/kWh for deep discharge lead acid batteries with 7 1/2-year life).

The most widely encountered applications are:

lighting by fluorescent tube
television sets and radios

refrigerators and freezers
fans and ventilators
transceivers.

The appliances adapted to these generators, in 24 or 12 V dc, exist today on the international market and provide satisfaction on the whole, although their distribution in the developing countries is still limited.

COSTS

Small photovoltaic generators cost between $13 and $20 per peak watt installed capacity depending on the size of the order, sales expenses, type of system, and so on. This represents, for instance, US$2,500 for a 150-W system for lighting (four

Table 21.1

Comparison Between Photovoltaic Lighting and Kerosene Lighting: Example 1—PV Compared to Ordinary Kerosene Lamps

Photovoltaic system(*)		Kerosene lamps(**)	
Number of lamps	2	Number of lamps:	3
Number of hours/day	6	Number of hours/day:	6
Fluorescent lamp capacity W	8	Consumption cm3/h/lamp:	20
Power in lumens (per lamp)		Power in lumens (per lamp)	
per lamp	400	per lamp	60
PV peak power W	40		
Total capital cost US$	640	Total capital cost US$:	20
System cost US$/Wp	16		
% cost modules	60		
% cost batteries	15		
% cost miscellaneous	25		
Module life years	15	Lamp life years:	2
Battery life years	5	Kerosene consumption	
		1/month:	10.8
Miscellaneous life years	10	Kerosene cost $/1:	1.00
Maintenance costs in		Maintenance cost in	
% capital cost	3	% capital cost:	15
Calculated annual cost		Calculated annual cost	
Depreciation in $	102	Depreciation in $:	12
Maintenance in $	19	Fuel in $:	130
		Maintenance $	3
Total annual cost in US$	121	Total annual cost in US$	145
Monthly cost in US$	10	Monthly cost in US$	12
% operating expenses	16	% operating expenses:	91

(*)The PV system offers better quality service: availability, rapid startup, etc.; operating costs are lower; initial capital investment will often necessitate a purchasing loan; interest rate is assumed at 10%.

(**)Ordinary lamps with low luminosity (60 lumens/lamp vs. 400 lumens/lamp for PV) are considered; the service offered is very mediocre in comparison with PV; the data obtained are from surveys in India, Indonesia, Sri Lanka, and West Africa.

to five lighting points) and to run a television set, or US$600 for a 35–40-W system, which is the minimum size for charging a 12V battery.

This system can already be more economic than the extension of a local grid if the user is located at a distance of more than 500 to 1000 m from the existing grid, and if user needs in energy remain limited (less than 500 Wh/day for example).

Tables 21.1, 21.2, and 21.3 provide a number of examples of comparative cost analyses in different contexts, compared with competing energy alternatives, such as a small diesel generating set, kerosene lamps, and such.

CASE STUDY

Photovoltaic battery chargers. Central to the diffusion of small battery charging photovoltaic systems is the provision of financing. A cooperative scheme in the Dominican Republic illustrates how this can be achieved.

Table 21.2
Comparison Between Photovoltaic Lighting and Kerosene Lighting: Example 2—PV
Compared to Coleman (pressurized kerosene) Lamps

PV (*)		Coleman (pressurized kerosene) lamps(**)	
Number of lamps	3	Number of lamps	3
Number of hours/day	6	Number of hours/day	6
Fluorescent lamp capacity W	13	Consumption cm^3/h/lamp	50
Power in lumens (per lamp)	800	Power in lumens (per lamp)	700
PV peak power W	100		
Total capital cost US$	1,600	Total capital cost US$	200
System cost US$/Wp	16		
% cost modules	60		
% cost batteries	15		
% cost miscellaneous	25		
Module life years	20	Lamp life years	5
Battery life years	5	Kerosene consumption	
		1/month	27
Miscellaneous life years	10	Kerosene cost $/1	0.65
Maintenance costs in		Maintenance cost in	
% capital cost	2	% capital cost	10
Calculated annual cost		Calculated annual cost	
Depreciation in $	241	Depreciation in $	53
Maintenance in $	19	Maintenance in $	20
		Fuel in $	216
Total annual cost in US$	260	Total annual cost in US$	289
Monthly cost in US$	23	Monthly cost in US$	24
% operating expenses	12	% operating expenses	82

(*)The PV system offers better quality service: availability, rapid startup, etc.; operating costs are lower; initial capital investment will often necessitate a purchasing loan; interest rate is assumed at 10%.

(**)Pressurized kerosene lamps are effective but delicate to use; the price of kerosene is variable and may exceed $3/1; the three-lamp solution applies to a comparatively affluent family.

Table 21.3
Economic Comparison Between a Diesel-Generating Set and a Photovoltaic System:
Example—Electrification of a Solar Lighthouse in Polynesia

Diesel-generating set(*)		Photovoltaic system(**)	
Number of lamps	4	Number of lamps	4
Number of hours/day	5	Number of hours/day	5
Lamp consumption in W	60	Lamp consumption in W	13
Other uses: Wh/day		Other uses: Wh/day	
(refrigerator, TV,		(refrigerator, TV,	
miscellaneous)	1,300	miscellaneous)	540
Capital in kW	3	Peak capacity Wp	288
Capital cost in FF	25,000	Capital cost in FF	43,200
Capital cost US$	3,330	Capital cost US$	5,760
Life-cycle in years	6	% cost modules	50
Running time h/day	6	% cost batteries	12
Consumption in 1/kWh	0.8	% cost miscellaneous	38
Fuel cost FF/1	6	Life-cycle modules years	15
Maintenance and parts		Life-cycle batteries	5
cost: % cap. cost	10	Life-cycle miscellaneous	10
Output Wh/day	25,000	Maintenance cost in	
		% cap. cost	2
Calculated annual cost		Calculated annual cost	
Depreciation in FF	5,740	Depreciation in FF	6,879
Fuel	4,380		
Maintenance in FF	2,500	Maintenance in FF	864
Other	500	Other	200
Total annual cost in FF	13,120	Total annual cost in FF	7,943
Total annual cost in US$	1,750	Total annual cost in US$	1,060
Monthly cost in FF	1,093	Monthly cost in FF	662
Monthly cost in US$	146	Monthly cost in US$	88
Percent of operating costs	56	Percent of operating costs	13

(*)The generating set here is underutilized, a situation that is frequently encountered. Electricity is not available round the clock, hence problems for the refrigerator. The service life of the set may vary substantially, depending on maintenance, make, etc. Rounded-off values. 1 US$ = 7.5 FF.

(**)Electricity is available round the clock. Recurrent costs and maintenance are very low. No fuel required. Services and not kW are provided here. The equipment must be 24 V. A 24Vdc/220 Vac converter can be used for small units. Interest rate i = 10%.

There are 540,000 rural homes in the Dominican Republic, of which 380,000 are without electricity. Typically these homes use kerosene and dry cells for lighting and rechargeable car batteries for television, according to the consumption and prices shown in Table 21.4.

This consumption could be largely replaced by a 10–15-W solar PV system for minimal lighting and radio or by a 35W system for television as well. The corresponding cost breakdown is given in Table 21.5, from which it can be seen that 20-year foreign exchange costs are $520 for kerosene and dry cells or $720 including car battery (for television). Against this we have a $345 foreign exchange cost for PV.

Table 21.4
System Foreign Exchange Costs (in U.S.$)*

Item	Price	Estimated foreign exchange component
One time:		
PV module 35W	$210.00	$210.00
control box	20.00	10.00
other hardware	50.00	25.00
		$245.00
Recurring:		
battery	40.00	10.00/2 years

*Dollar costs of PV are U.S. $245 (one time) plus U.S.$10 × 10 = U.S. $345 for a 20-year period, assuming no inflation of battery cost.

Table 21.5
Foreign Exchange Costs of Conventional Energy (in U.S.$)

Energy source	Typical consumption	Price	Estimated foreign exchange component	Yearly total
Kerosene	12 gallons/year	$ 2.00/gal	$ 1.00/gal	$12.00
Dry cells	96 cells/year	$ .30/cell	$ .15/ea.	$14.40
Car battery	1 battery/year	$40.00 each	$10.00 ea	$10.00

In practice, 35–40-W PV systems can be sold for $500 including distribution costs and installation. It was determined that with 10% downpayment and repayment at 15% interest over a 7-year period the number of rural households that could afford a 35-W PV system was about 66,000 or about 20% of those now without central electric power. The smaller 15-W PV systems, costing about $200 and sufficient for minimal lighting and radio only, could be afforded by the majority of the rural population if suitable financing were available. Starting with a rural cooperative association and $100,000 revolving fund, it was estimated that some 1,500 35-W PV systems could be installed over a 10-year-period.

Complementing the revolving fund is the requirement for a service center enterprise to install and maintain the systems. This would service two or three rural cooperative associations and could be financed by a $75,000 bank loan of which about a third would cover building, truck, two motorcycles (for technicians), and tools; a third for accounts receivable (1-year financing of 10 systems a month); and a third for an inventory of 100 panels and parts. The scheme has been initiated with grants totalling only $20,000 but already 100 homes have them installed, and with additional funding the scheme could expand rapidly.

ADVANTAGES AND DISADVANTAGES

Small solar generators for domestic use offer the following advantages:

1. Simple installation and use.
2. No moving parts, reliability.
3. Electricity available around the clock, in contrast to small private diesel-generating sets.
4. Reasonable capital cost, often cheaper for the community (very high cost distribution grids in sparsely populated areas).
5. Broad technological transfer possible; including local manufacture of module supports frames, regulators, batteries, refrigerators.
6. Low operating costs.
7. Control of the installation by the beneficiary.
8. Substantial improvement in living conditions.
9. Interesting solution in certain contexts for rural pre-electrification (5 to 10 years before the public electric grid).

The disadvantages include the following:

1. Initial cost still high for the modest user compared with standard alternatives and frequent need to secure a loan for the initial equipment at preferential conditions (e.g., Polynesia).
2. Unsuitable for large energy requirements (air conditioners).
3. Need for specific user equipment adapted to 12 to 24 V dc, which is less readily available than 220 or 120 V ac.
4. Low flexibility in case of sharp increase in requirements (high marginal kWh cost).
5. Individual equipment maintenance must be set up.
6. Battery replacement every 3 to 5 years.

PRESENT MARKET AND DEVELOPMENT CONDITIONS

The use of these small PV generators is growing in several developing countries, particularly in Polynesia, Papua New Guinea, and Zimbabwe. The total number of installations runs into a few thousand. A significant potential market exists in many countries, where some users are ready to acquire such PV generators for domestic lighting, television, refrigerators, and the like.

To achieve widespread dissemination and to meet the needs of potential users economically, it will be necessary to:

Develop very simple, reliable kits produced cheaply due to the scale effect.

Develop adapted "financial products" for initial funding as has happened in parallel cases in agriculture.

Train the users and maintenance personnel.

Find appropriate distribution channels.

Set up demonstration operations.

CONCLUSIONS

In some contexts, low capacity solar photovoltaic generators can contribute economically in meeting priority needs such as lighting, food storage, education. This proven technology, which is sometimes less expensive for the community than conventional rural electrification, should be considered systematically as a possible alternative in development projects. In a number of countries it will enable pre-electrification for the scattered rural populations in the next 20 years. In the first phase, the equipment of these populations should focus on collective uses, such as schools, meeting halls, dispensaries, businesses, public lighting.

Beyond the "institutional pre-electrification" stage, major possibilities appear to exist in certain countries for "family" generators, which could be disseminated by a number of channels, chiefly private. An institutional impetus may, however, be useful to facilitate the opening of this market: tax exemption, credits, information, training, and such.

22

Renewable Energies and Rural Telecommunications

DESCRIPTION

Telecommunications are of considerable economic importance in development. All too often they are financed as a consequence of economic development; however, the International Telecommunications Union (and others) have indicated through well-documented case studies that rural telecommunications can so facilitate efficient development that they should be implemented as a prerequisite for development, rather than as an afterthought.

The main economic spin-off from efficient telecommunications is through the reduction of unnecessary travel and unnecessary deliveries of materials and the improvement of logistics needed to permit economic activities in the rural areas. The management of the rural economy can obviously be improved through reliable communications. Public telecommunications are also widely perceived as desirable, even by relatively poor people, and hence an efficient telephone network improves the sense of well-being of rural populations and can therefore be of political importance. Finally, a rural telephone system, providing it functions reliably and efficiently, has the potential for being a profitable economic activity in its own right and should not require subsidies (even though they may be justifiable due to the other spin-offs that may be gained).

Unfortunately, most rural telephone networks in developing countries are at present extremely limited and highly unreliable. Typically, in African countries over 75% of telephones are in the urban areas, whereas over 75% of people live in the rural areas. The rural network will generally either be unserviceable or, even when it works, severe traffic congestion and poor signal quality will make its use unattractive and cause people to seek other methods of communicating, often at much greater cost to both themselves and the local economy. Hence existing rural telephone networks often are so unsuccessful as to be net loss-makers as revenue fails to balance costs.

The main constraint that has led to this situation has been difficulty in providing reliable power supplies for rural telephones and exchanges. However, the use of solar photovoltaics combined with improved, more cost-effective, and lower powered electronic telecommunications equipment can revolutionize this situation by, for the first time, offering a real practical possibility of reliable rural telecommunications for general use.

In the future, the use of geo-stationery telecommunications satellites promises even further dramatic improvements by connecting the remotest corners of large countries with high quality, but low-powered telephone connections. This will, of course, also demand small autonomous power sources to power the ground stations.

STATE-OF-THE-ART

There are three main areas requiring power supplies in order to establish a reliable and modern rural telecommunications network:

1. Trunk line repeater stations (along the backbone).
2. Manual or automatic telephone exchanges (network modes).
3. Single channel spurs to rural terminals.

All of these must function effectively or trunk calls from rural terminals will fail to reach their destination. At present most trunk lines use multichannel UHF often with repeater stations every 30–80 km. Such repeater stations usually run the telecommunications equipment off batteries that are recharged every day by running one of a pair of diesel-generating sets for up to 10 hours (two engines are essential to provide standby capacity). Because most PTT authorities procure equipment by competitive bidding, there is a common tendency for the battery bank to be undersized (manufacturers are tempted to do this to keep costs down in order to win bids) and as a result the systems are generally inefficient in matching generating sets to the load (so heavier fuel consumption than necessary occurs) and battery failure occurs sooner than necessary.

Exchanges, usually in provincial towns, generally use similar facilities to repeater stations, although where provincial towns have main power, this is normally used to recharge batteries. In most situations even with mains power, standby generators are provided in case of mains failures. Multichannel repeaters and telephone exchanges typically have electrical energy needs in the region of 3–10 kWh per day using modern equipment.

Spurs to provide telephones in remote villages and settlements are normally single channel, VHF or UHF, line of sight from the exchange or from a repeater on the trunk route. Where distances are too great or the terrain is too uneven for line-of-sight routing, then single-channel repeaters need to be installed as necessary. The only generally practical means to provide power for a single channel VHF/UHF to a remote settlement is using solar PV cells (or a small wind generator) to recharge a battery. In some cases standby generators, usually provided for some other purpose, or batteries recharged in a local town and regularly exchanged have been used. However, the organizational and logistical problems usually cause eventual discharge of the battery and system failure unless an automatic recharge is provided with PV cells.

Power requirements for a single channel telephone transceiver and telephone unit are in the region of 50–250 Wh/day depending on the amount of telephone traffic and the transmitting power necessary, so commonly a single solar PV module is sufficient. Both the cost and the power requirements for VHF/UHF telephone systems are declining due to improvements in telecommunications technology and greater competitiveness between suppliers in this sector.

SPECIFIC CONSTRAINTS

There is a commonly held perception that rural telecommunications are neither technically feasible nor economic. However, recent developments in solar and wind power generation for small-scale applications coupled with improved lower powered electronic, high-frequency telecommunications systems make this an outdated view. Nevertheless the provision of rural telecommunications is rarely considered or even when it is, it tends to be given a low priority. The trunk and urban network with which any rural system interfaces requires adequate capacity to handle the extra traffic.

The PTT authority requires the staff and infrastructural capability to service and maintain an effective rural operation. There is a definite problem with the siting of rural public call boxes so that they are conveniently located to attract users, and so that they can be kept under observation by a caretaker who should report faults, keep the system clean, and possibly help to educate would-be users who are not familiar with how to use a telephone. One solution being explored in the Gambia is to locate rural public pay phones alongside or within rural health centers and to make health center staff responsible for overseeing the unit. Health center paramedics are generally interested in having a working telephone close at hand and are therefore readily prepared to cooperate in such an arrangement, especially if offered an incentive such as a certain number of free telephone calls.

COSTS

A typical single channel VHF public telephone (with booth, coin collector unit, mast, and PV power supply plus the necessary transceiver and antenna), which interfaces with the nearest town exchange, costs in the region of US$5,000 to $10,000 of which about 10–20% covers a solar (or wind) power system and battery.

Typically, it is only necessary to generate 30–60 minutes of traffic per day to make a single channel VHF spur break-even under commonly used telephone tariffs. Experience has indicated that, even in poor communities, once a reliable system has been established (so that people expect calls to be successful), a profitable level of traffic yielding a good return on the investment may commonly be generated. The results are obviously fairly location-specific and therefore individual surveys are needed before planning any such investment (see Table 22.2).

If a repeater station is needed due to the location not being on a "line-of-sight" from the exchange, this will cost in the region of an extra $5,000 including its power system, and obviously more traffic will be needed to break-even.

Telephone exchange and multichannel trunkline repeaters may also advantageously be solar powered. Various studies have indicated that whereas solar (and in some cases wind) power is likely to offer a primary benefit of reducing system downtime due to power failure (engine generators are less reliable than solar/wind systems), in many cases the actual unit electricity costs of generating the necessary power are also reduced. Without question, solar- or wind-powered systems will have a major impact in reducing recurrent costs and in freeing technicians and engineers to concentrate on the telecommunications equipment rather than having to maintain and repair engines. Fuel distribution and storage problems, which can be significant in some countries, are also removed. A comparison of the present costs of diesel-powered repeater stations in the Gambia, West Africa, with a replacement solar photovoltaic power system is presented in Table 22.1. It can be seen that the solar option is significantly cheaper.

POTENTIAL FOR LARGE-SCALE DEVELOPMENT

The scope for converting existing rural telephone systems to solar or wind power is quite large, but there is also a huge possibility for extending the rural telephone

Table 22.1
Conversion of UHF Repeater Stations in the Gambia

Load 4.25 kWh/day	Solar costs ($)	Actual diesel costs ($)
Power system capital costs	18,238 to 27,220	6,463
Annual O & M costs	100 to 200	3,938
Annualized life-cycle costs	2,617 to 3,869	4,914
Unit electricity costs/kWh	1.68 to 2.49	3.17
Cost as % of diesel	53% to 78%	100%

Table 22.2
VHF Single Channel Links in the Gambia

Capital cost per link	$6,700 to 9,000
Break-even traffic required	26 to 34 minutes/day
Power system cost	$1,000 to $2,000

network in Africa, Asia, and Latin America, and the economic and development implications of doing this could be very significant.

With falling costs and constant improvements in the state-of-the-art, both for power systems and for telecommunications equipment, the cost-effectiveness of rural telecommunications may be expected to improve further and thereby to increase the scope for further investment in this sector.

23

Renewable Energies and Health Centers

ENERGY REQUIREMENTS

The energy requirements of rural health centers depend on such factors as the size and number of people served. The main requirements are for lighting and refrigeration. Lighting is reported on the earlier; this section deals only with refrigeration.

The energy required to operate a refrigerator depends on its loading—in particular the amount of ice produced and the ambient temperature. A typical electrical requirement is between 0.5 and 1.5 kWh/day for a medical refrigerator producing 1 kg of ice per day with 32°C ambient temperature.

This is a very small requirement. Together with lights it is much less than would be necessary to consider using an engine/generator. Typically off-grid sites use kerosene-fueled absorption cycle refrigerators, requiring around 0.5 to 2.0 liters/day of fuel (lighting is also generally by kerosene).

STATE-OF-THE-ART

Generally there is no electricity in rural areas where these hospitals and health centers are located, or at best fuel and power supplies are erratic and unreliable. This problem will grow as health services are extended into more remote and previously unserved areas. Refrigerators are essential equipment in the vaccine cold chain. Solar-powered refrigerators are an important potential solution to this problem and are often claimed to have better performance, lower running costs, better reliability, and longer working life than kerosene or bottled gas refrigerators, or diesel generators.

Photovoltaic-powered refrigerator systems that have been specifically designed for this purpose are being aggressively marketed by more than 40 suppliers, and the number is growing. Sample refrigerator systems have now reached more than half the developing regions of the world and most are being informally evaluated by health managers, mission health staff, and health workers. More than 800 installations are known.

A photovoltaic array charges a battery via a voltage regulator, which is employed to prevent the battery from being overcharged. The battery powers a dc motor, which is coupled directly to the compressor. The motor/compressor is usually manufactured as a hermetically sealed unit. The motor is of the electronically commutated, brushless type. A second regulator is employed to ensure that the motor/compressor is run only within its rated power range and to prevent overdischarge of the battery. Freon refrigerant is used in the cooling cycle in the normal way. A thermostat is used to switch the compressor on and off as required. The basic WHO specification for photovoltaic refrigerators is given in Table 23.1.

CONSTRAINTS TO WIDESPREAD DIFFUSION

As noted above, there have been many demonstrations of photovoltaic medical refrigerators. Unfortunately, little field performance data have been produced, analyzed, or disseminated. In addition many failures have been reported in a "casual" manner (without supporting technical details of the mode of failure) and many people, including policy and decision makers in funding and purchasing agencies, are skeptical about the technology. Nevertheless the PV industry clearly believes there is a market and, for the present at least, is committed to developing it. There is a crucial need to assess the technical and cost status of the present generation of PV refrigerators and give guidance to potential users and manufacturers.

The widespread diffusion of photovoltaic medical refrigerators may be constrained by the development and use of vaccines and medicines that do not require refrigeration. However, this cannot occur in the immediate future. It is believed that immediate development and exploitation of the medical market will in fact lead to the use of PV refrigerators for other purposes, for example, cold drinks and food preservation. This "new" market is orders of magnitude larger than the medical market and should be of great importance to the PV industry.

Table 23.1
WHO 1981 Outline Specification for Photovoltaic Refrigerator

Net vaccine capacity	30–40 liters (top opening)
Ice-making performance	Minimum 1 kg/24 hrs in 32°C ambient
Refrigerator performance	No part of the vaccine storage area to exceed 8°C or drop below −3°C in: 43°C ambient temp / 32°C ambient temp / 43°C day time/15°C night time
Holdover time	More than 6 hours below 10°C when power cut out in 43°C outside temp
External casing	Noncorrodable
Minimum battery maintenance interval	1 year
Insulation	Rigid polyurethane

ADVANTAGES AND DISADVANTAGES

The main advantages of solar-powered refrigerators are:

1. Better performance and reliability than kerosene refrigerators.
2. No fuel requirement.
3. Longer life.
4. Better temperature regulation than kerosene.

The main disadvantages of solar refrigerators are:

1. High capital cost.
2. Limited field experience.
3. Current modules have limited ice-making capability.

COMPARATIVE COSTS OF KEROSENE AND PV REFRIGERATORS

Analyses of the installation and operating costs of photovoltaic and kerosene refrigerators have only compared the PV system installation cost with the savings from avoided-kerosene purchases. This is not a realistic approach. Medical care and immunization programs are not "economic activities." Meaningful quantifiable results relate to the *relative costs of the options* and also their likely influence on *achievement of immunization program goals*. This latter point is particularly important since the fixed overheads for any medical program are generally large compared with direct vaccine refrigeration costs, and hence only by taking these costs into consideration can the financial benefits of PV refrigerators become apparent.

The financial benefits to be expected from the use of solar vaccine refrigerators may be summarized as follows:

1. More doses of vaccine successfully delivered per unit of money invested because PV refrigerators are more reliable than kerosene.
2. Savings from kerosene substitution.
3. Savings from eliminating the need for paying for kerosene transportation.
4. Savings from reduced maintenance and wages.
5. Savings by reducing the requirement for back-up transporting of vaccines due to unservicable refrigerators.

Improved refrigerator reliability will increase the effectiveness of immunization both with initial and follow-up "booster" shots.

Kerosene-fueled refrigerators used in the vaccine cold chain have an initial capital cost of only $300 to $800, considerably less than a PV refrigerator. With transportation and installation this may rise to $1,500 installed compared with typically $6,000 for an installed PV refrigerator. The operation and maintenance costs are, however, very large and their reliability very low. The World Health Organization has estimated that 71% of the life-cycle costs are for fuel and maintenance. In many countries the cost of kerosene at the point of use can be as high as $2 per liter but typically $0.70/liter. Hence the fuel costs of a kerosene refrigerator consuming 1 liter/day are typically $0.70 per day.

The availability of kerosene refrigerators in field use, that is, the percentage time the vaccines are maintained at correct temperatures is on average only 50% because of fuel and spare part supply problems. This poor reliability of kerosene refrigerators compared with PV refrigerators affects the comparative costs significantly.

A methodology used for financial/economic analysis consists of calculating life-cycle costs for each option by taking the summed present values of their respective cash flows. This can then be annualized and divided by the number of vaccinations per year to give the cost per dose. An alternative would be to divide the annualized cost by the refrigerator capacity *and availability* to give the cost per liter-month of capacity. The assumption given in Table 23.2 shows that solar refrigerators can typically be the least costly solution.

The cost per dose comprises a direct cost due to refrigeration plus a cost due to the program overheads. Hence the better the productivity (more reliable the refrigerators), the lower the overhead cost per dose. Such an analysis recently undertaken on an immunization program in the Gambia concluded that the overhead cost per dose is reduced by 6 to 7 cents by using a PV refrigerator because of the better reliability. Thus the overhead finance is used more effectively. The overall cost per dose is cheaper for the PV refrigerator even where the PV refrigerator capital cost is high.

POTENTIAL FOR LARGE-SCALE DEPLOYMENT

Quantitative data on refrigerator requirements for health centers in the WHO immunization program are available. If 50% of all new health centers were equipped with solar refrigerators, this would involve between 500 and 1,500 units per year.

Table 23.2
Data for Comparative Cost Analysis

| Parameter | Photovoltaic | | Kerosene | | |
	A*	s†	low case	typical	high case
Net vaccine capacity (liters)	100	100	100	100	100
Initial capital cost ($)	5200	4500	300	500	1000
CIF and installation ($)	1800	1500	500	000	1000
Fuel costs ($/day)	—	—	0.25	0.70	3.00
Maintenance costs ($)/year	150	150	50	100	150
Lifetime (years)	15	15	10	5	2
Availability (% time in service)	95	95	80	50	20

*System sized for 3.5 kWh/m2/day worst month insolation.
†System sized for 5.8 kWh/m2/day worst month insolation.

However, the WHO Expanded Program on Immunization (EPI) has a goal of providing immunization facilities to all the world's population by 1990. Although this is not expected to be achieved (current coverage is 50% compared to 5% at the formation of EPI), WHO and UNICEF have embarked on an accelerated immunization activity that calls for several thousand refrigerators for cold chains per year. As it is important to provide *sustainable* immunization facilities, PV refrigerators are increasingly being specified.

ACTIONS REQUIRED FOR LARGE-SCALE UTILIZATIONS

There is an immediate need to prove beyond doubt that the use of photovoltaic refrigerators can make a significant and cost-effective contribution to improved health care. To achieve this requires the following:

1. Careful selection of the region/country to be served—taking account of health department interest, needs, accessibility, communications, maintenance support available, etc.
2. Adequate design of the project—selection of sites, sufficient budget for technical assistance, repairs/replacements, etc.
3. Initial training for users and of technicians for monitoring and maintaining the equipment.
4. Provision of technical assistance to the health service agency providing the refrigerators, including ongoing reinforcement training (local staff often change).
5. Confirmation of true operating costs of kerosene refrigerators such that comparative PV kerosene refrigerator economic studies can be completed.
6. Up to 50% of the cost of a solar refrigerator is for the PV. Improved compressors with much higher efficiencies than the present 50% will permit correspondingly less PV, the cost of which will in any event decline, so that substantial cost reductions can be expected.

24

Strategies for Rural Electrification

PRESENT CONSTRAINTS TO THE DEVELOPMENT OF RENEWABLE ENERGIES

With few exceptions today in the Third World, renewable energy does not play a significant role in the decentralized generation of electricity. The extension of interconnected grid remains the sole means of supplying electricity to the rural populations in a significant manner.

The really noteworthy exception in China, which intensively developed micro-hydro power stations (about 90,000 power stations representing an installed capacity of 6,600 MW) and which has also developed the production of biogas, a small part of which is used in internal combustion engines to generate electricity. The "dendrothermal" (wood-fired) power stations developed significantly in the Philippines are connected to the interconnected grid and hence are not included in the decentralized generation of electrical energy in rural areas.

In nearly all the developing countries, decentralized electricity generation projects have been implemented or are under examination. However, these are localized, without any real impact on the standard of living of the rural population or on the national energy balances. They do not really offer valid alternatives to electrification by grid extension.

A few years ago, many Third World countries expected a "revolution" in the energy supply of the rural populations, thanks to renewable energy. This revolution has failed to occur, and the obstacles that faced and are still facing its widespread diffusion in the rural environment can be identified.

1. First, there is cost. Preceding chapters show that the initial capital cost in renewable energy hardware is still high, indeed very high, and nearly always higher than the cost of conventional alternatives rendering equivalent services. We have also seen that the operating costs are not negligeable.

   The predictions made by experts and industrialists a few years ago on the rapid drop in the costs of renewable energies, especially photovoltaic conver-

211

sion, have not been borne out by the the facts. Many uncertainties also prevail concerning the real cost of decentralized electrical energy generation, both by conventional methods with internal combustion engines and by renewable energy systems. The real life cycles, maintenance costs, and such are still relatively unclear for both alternatives. Despite these uncertainties, it can be asserted that although the initial capital cost is high, several renewable energy systems do sustain an economic comparison today with conventional alternatives, at least in certain conditions.

However, even if the comparison is favorable to the renewable energies, the high capital costs pose a serious obstacle to product dissemination. In many Third World areas, rural monetary incomes are very low. The purchase price of renewable energy equipment discourages households and local communities, and very few appropriate loan systems have been set up to surmount this handicap.

2. The questionable image of renewable sources of energy and information about them. A great deal was expected from the renewable energies. A large number of so-called demonstration operations have been funded by several aid agencies in recent years. But many of them dealt with new products that were not yet ready for marketing, and they often turned into a "counterdemonstration."

   The image of renewable energies (or at least of some systems) has accordingly become, for several developing country governments and for many aid agencies, that of techniques not yet mature or not yet adapted to the conditions prevailing in the Third World.

   Some systems retain a favorable image, such as micro-hydro power stations. But whatever the image, the information of the policymakers, both in the northern and the southern countries, is generally inadequate. This information about renewable energies is often presented in a partial manner, with very little reliable data emanating from independent sources. Hence very few decision makers possess a sound vision of the possibilities, implementation constraints, costs, or prospects of each system.

3. Adaptation to needs and to constraints. The development of renewable energy products has not hitherto paid due consideration to the specificity of needs and constraints (and also their very wide diversity) in the Third World. The products are often more the result of technological research conducted in a developed country than a response to a clearly identified need in a developing country, taking account of local constraints.

   Research centers, including those that were set up by the developing countries themselves, have not really succeeded in establishing a link between the development of technologies in the laboratory, industrial fabrication, and the real needs of the rural world.

   The consequence of these obstacles is that in most of the Third World today, there does not exist a genuine market for renewable energies, nor a valid alternative to conventional rural electrification. Only localized projects exist, and these are often implemented thanks only to funding by certain aid agencies.

Yet other agencies, their fingers burned by disappointing experiences, now demonstrate very little enthusiasm for the funding of new renewable energy projects.

POSSIBLE TECHNOLOGIES

Does this mean that in most of the third world today there are no possibilities of developing a market for renewable energies in the rural areas? The technical and economic details that have been compiled in the preceding chapters show that in the immediate future and in the medium term real possibilities do exist of producing decentralized energy from renewable sources.

1. *Wind generators*. These are a competitive source of electricity in windy areas and are susceptible to substantial development in many Third World areas. The implementation constraints do not appear prohibitive. However, they have only just begun to spread in the industrial countries (California, Denmark, etc.) with very few projects in the developing countries.

   The experience we have today is essentially with wind generators supplying the interconnected grid, and data are still unavailable on wind generators supplying mini- or micro-grids in competition with conventional generating sets. Subject to this reservation, this system appears to deserve closer scrutiny.

2. *Micro-hydro power stations*. Micro-hydro power stations have begun to spread in several Third World countries. A number of projects are really successful and supply rural communities with cheaper electric power than any other source. Others have proved to be costly and sometimes not reliable enough. Two problems need to be solved: the problem of *capital cost,* which must be kept as low as possible by the use of standardized hardware and local engineering, and the problem of the *grid load factor,* which must be high enough to make the project viable. This may require the establishment of new activities consuming energy permanently. If these solutions can be implemented, the system boasts major development prospects.

3. *Wood-fired gasifier micro-power stations*. These seem capable of supplying a kWh at a price comparable to that of a conventional generating set under certain conditions: isolation of the locality to be supplied and availability of raw materials, with the advantage of using a local energy source. The problems to be solved are those of the management of the resource to guarantee regular supply and that of maintenance.

4. *Wind pumping*. Modern wind pumping units are widespread throughout the world, in Australia, the United States, Argentina, and China. On the contrary, they are virtually absent from other Third World areas, especially in Africa, where they could render real services in supplying water to livestock and humans, and possibly for irrigation.

   In these areas, the development of this system is faced with the lack of local capacity to determine the equipment to be set up, to install it, and to maintain it. Development is also hampered by the high cost of imported units. The future trend will certainly be to develop local engineering capacity and local manufacture of sufficiently effective and reliable hardware.

5. *Photovoltaic pumping*. A large number of photovoltaic pumps now exist, in service in the Third World countries, to overall satisfaction. The capital cost of these units is still high, but in areas where wind pumping is inadequate, this system offers a competitive solution if mechanical pumping is necessary (which is not always the case; a hand pump is often adequate), and if the capacity required is low (this often applies to pumping for human consumption and more rarely to pumping for irrigation).

6. *Photovoltaic equipment for individual and collective needs*. Preliminary experience is available with small photovoltaic generators supplying electricity to meet basic rural needs. This system offers the rural populations the benefit of certain services rendered by electricity (lighting, refrigeration, television, etc.) if no other process proves viable. These units incur low operating costs, and the main problem is still their relatively high purchase price. The large quantity manufacture of simple, reliable kits and the setting up of credit systems should help to solve this problem.

Micro-hydro power stations and wind generators and possibly wood-fired gasifier power stations offer alternatives to conventional rural electrification in certain conditions, alternatives that serve to supply a small electricity distribution grid.

The other systems cannot really substitute for conventional rural electrification. They enable the populations to enjoy certain services rendered by electricity. Small photovoltaic units in particular allow pre-electrification of the villages while awaiting the development of a grid, whether interconnected or not.

PREREQUISITES TO THE OPENING OF MARKETS

The example of a number of developing countries, particularly China, shows that these different possibilities can be implemented and that renewable energies can be developed to make a significant contribution to the energy supply of the rural world. The example of a number of other countries also shows that the development of renewable energies in the rural world will not take place spontaneously and that voluntarist development policies for renewable energies will not necessarily culminate in concrete results.

Under what conditions can the renewable energies be developed in a country to provide an alternative to conventional electrification methods? It is clear that continued research and development are necessary to design cheaper, more reliable equipment, better adapted to the conditions prevailing in the Third World. However, the experience of recent years also shows that this is not sufficient, and this experience suggests that certain conditions must be satisfied for the widespread diffusion of renewable energies to be achieved successfully.

1. The existence of political will. It seems clear that the diffusion of renewable energies in rural areas will not take place by counting on the commercial sector exclusively, if only because in all the countries of the world, rural incomes depend closely on the policies implemented by the authorities. In all the world's countries, the government also plays a key role in energy supply, particularly by its pricing policy.

   Many developing country governments proclaim their interest in the development of renewable energies. Yet the real interest may be quite different from one country to another. It may be limited to support for local research and development institutions, or to the search for project funding, from aid agencies. It is improbable that such policies will lead to the widespread diffusion of renewable energies that change the standards of living of the rural inhabitants.

   On the other hand, the implementation of systematic development policies for renewable energies in a sector, such as telecommunications and health, can certainly enable the initial diffusion of these energies. Yet neither the governments nor the aid agencies can themselves achieve a widespread diffusion bringing a change in the standards of living of the rural populations. By contrast, the governments, possibly backed by aid agencies, are the only entities capable of creating conditions that are favorable to widespread commercial diffusion, by developing and implementing appropriate policies.

2. The existence and knowledge of resources. Real renewable energy resources are generally poorly known. In a given area, there rarely exists more than one renewable energy source, that is technically and economically usable, and priority must be assigned to the development of the system that harnesses it. What is this source? What are its characteristics? It is indispensable to be able to answer these questions before any likelihood of promoting renewable energies in the area. Nor is it necessary to compile exhaustive data about the resource in order to implement a renewable energy development policy. Excessive detail in analyzing the resource could be a pretext to postpone action indefinitely.

3. The creation of local technical capacities. The major handicap to practically all renewable energy systems is the high initial capital cost. It is obvious that if the technology is imported from the industrial countries, the chances of having competitive equipment are very slight. The widespread diffusion of renewable energies is only conceivable if at least part of the technology is transferred locally. In practice, this can be achieved by: creating local engineering capacities that can design and prepare the setting up of the equipment; locally establishing at least part of the manufacture; and creating maintenance capacities. This implies a major reorientation of the present infrastructure, which consists largely of: many research centers in the Third World, which conduct academic research, without any serious impact on the rural world; and industries in the developed countries, which produce expensive equipment that is not necessarily ideal for the needs of the rural populations of the developing countries.

4. The creation of an appropriate funding system. Renewable energies cannot pos-

sibly be developed significantly without acceptable funding of the initial capital cost. This implies the existence of credit institutions that can handle the funding of the equipment purchased by the households or communities. In some cases, it is uncertain whether the financing institutions are capable of playing a major role in this area, and it may be indispensable to set up intermediate entities— mutually guaranteed rural associations, for example.

SUGGESTIONS FOR ACTION

A few main strategic guidelines can be suggested:

1. To improve the quality and dissemination of information about the possibilities of renewable energies in rural electrification. This involves familiarizing the public with the real costs and performance of renewable energies and conventional alternatives, the successes and failures of projects and policies and their causes, and making sure that this information reaches the policymakers. An organization of network generating and disseminating unbiased information is lacking today.
2. To insert renewable energies into the sectorial policies of the authorities, in areas of health, telecommunications, and so on.
3. To set up the conditions of the widespread diffusion of renewable energies in the rural environment, and to establish guidelines to:
   Encourage the creation of products adapted to needs by the drafting of specifications, the delivery of quality labels
   Inform the population about the possibilities and costs
   Favor the creation of engineering and manufacturing capacities, at least partial, on the spot
   Set up the necessary funding institutions
   Plan maintenance by training artisans or by setting up adequate organizations.

Appendix: Life-Cycle Costing Procedures

1. For each technology, data must be collected on:

Site and nature of the technology for the application.
Capital cost (installed).
Annual operations and maintenance cost, often expressed as a percentage of capital cost.
Output, on a daily or annual basis (in kWh, or m^3 of water, etc.). This may be based on an assumed capacity factor, or on incoming resource and conversion efficiencies.
System life and the nature and frequency of any parts replacement.

Salvage value, if any, and when such salvage takes place.
Fuel cost and fuel efficiency (for conventional systems).
Discount rate and project life.

2. Calculations are made on:

(a) *Future equipment cost discounted to the present.* Future equipment costs are seen as less of a burden than "up-front" initial costs. A discount rate is used to find the "present value" of future costs. All costs—initial costs and the present value of future costs—are added to get the total "net present value." The formula for calculating present value is:

$$P = F \frac{1}{(1 + i)^n}$$

Where:

P = Present value of future cost
F = Future cost
i = Discount rate
n = Period, years

In this manner, total present values are found for capital cost, replacement costs, and salvage values.
Values of $1/(1 + i)^n$ are given in tables of discount factors.

(b) *All costs are annualized.* Annual maintenance and annual fuel costs (if any) are already on an annual basis. However, the net present value of initial cost, replacement costs, and salvage values must be converted to annual payments. The capital recovery formula is used to convert present values to annual costs:

$$A = P \frac{i (1 + i)^n}{(1 + i)^n - 1}$$

Where:

P = Present value, $
A = Annual costs
i = Discount rate
n = Project life, years

Values of $(1 + i)^n - 1$ are given in tables of annuity factors $i(1 + i)^n$ (reciprocal of capital recovery factors).

The *total annualized cost* can be found by adding:
Annual maintenance
Annual operations (fuel)
Annualized initial cost
Annualized salvage values
Annualized replacement costs

(c) *Annualized unit costs are found.* To different technologies, we compare the unit cost of energy output for each. This unit cost is found by dividing the total annualized cost (above) by annual output.

Discount rate and project life must be consistent between technologies to be compared. Different approaches to economic and financial analyses use different values for the discount rate. In some cases, the general market interest rate is used. Corporations may use the rate of return they expect on their investments. Governments often publish values to be used for their economic studies. In this case we assume we are addressing general government policymaking. A value of 10%, which is typically used by the World Bank and other institutions for general studies, is used here. Project life is assumed to be 20 years, as this is a typically assumed life for various renewable energy equipment.

Part IV

Application of Microcomputer Technology in Energy Planning

25

The Changing Role of Energy Planning

RECENT TRENDS IN ENERGY PLANNING

In the immediate aftermath of the oil price shocks of the 1970s, the focus of energy planning in most oil-importing developing countries, and indeed the reason for its establishment, was the concern over the impact of higher oil prices on the balance of payments. Most of the initial efforts centered on increasing the efficiency of energy utilization and developing any domestic fossil fuel resources as a means to reducing dependence on oil imports.

By the early 1980s, however, several other priorities began to emerge. For one, the problems of deforestation and fuelwood use reached the forefront of public policy debate in a large number of tropical and subtropical countries, which underscored the relationship between energy and the broader aspects of rural development and agricultural policies.

Equally serious was the emerging deterioration of the international financial environment: with many countries having to renegotiate immediate debt service obligations, potential sources of foreign capital began to be much more conservative in light of doubts about the medium to long-term outlook for the balance of payments of many of the poorer developing countries as well. In short, the problems of capital mobilization made investment planning much more difficult. In any event, better coordination among donors and the international financial institutions (IFIs) forced a fundamental change in the relationship between donors and the capital-intensive institutions of the energy sector (electric utilities in particular; o.g., in Sri Lanka in the period 1981–1986, the electric scotor is expected to account for some 37% of all public sector investment): even agreements with the IMF over short-term balance of payments assistance were made contingent on reforms in energy pricing, to say nothing of conditions imposed by the World Bank to raise tariffs to establish minimal self-financing targets.[1]

Whereas the recent decline in the international oil price has provided many oil-

---

[1]For further discussion, see, e.g., P. Meier and R. Chatterjee, "Electric Utility Planning in Developing Countries: A Review of Issues and Analytical Methods," National Resources Forum, July 1987.

importing countries with a short-term foreign exchange windfall, for many developing countries the implications are unclear, especially those that were able to offset the costs of higher oil prices in the period 1974–1982 by remittances from their workers in the booming Persian Gulf economies. But even where this is not a factor, energy planning must deal with a whole new set of issues associated with the oil prices. The duration of relatively depressed oil prices and the rate at which prices may rise in the 1990s pose very difficult issues for the long-term security of energy supplies and the continued need for investment to develop domestic resources. Indeed, how to deal with uncertainties has become one of the most pressing analytical problems in energy analysis, and one dictated not by the level of the oil price, but by its volatility.

In short, despite the temporary windfall of sharply lower oil prices in 1986–1987, energy planning faces new and difficult challenges. Capital mobilization and deforestation are fundamental issues not likely to vanish quickly. Energy planners must now deal with a much broader set of development and macro-financial constraints than previously, and investment planning in particular requires a much more sophisticated approach.

NEED FOR IMPROVED ANALYTICAL TOOLS

With more complex issues to be analyzed also comes the need for improved analytical approaches. This need has, of course, been recognized both by national governments and their newly created energy planning entities, as well as by multi-and bilateral assistance agencies that over the past 10 years have provided guidance and technical assistance to such bodies.

At the same time, the past few years have seen rapid developments in computer hardware and software technology to the point at which desk-top microcomputers have the computational power of the mainframes typical of the mid 1970s. (The new machines based on the INTEL 80386 chip bring vast now capabilities to low-cost, desk-top microcomputers, especially in graphics, adressable memory, multitasking, and speed). Thus microcomputers have become a natural adjunct to energy analysis since the types of analysis required are almost ideally suited to the microcomputer scale. For example, the data bases necessary for energy planning fit well into the 20–30 megabyte capacity of hard disks that populate the current generation of microcomputers. There is certainly no need for data bases in the gigabyte range, which are still the domain of mainframe and large minicomputers. Indeed, almost all national energy planning entities in developing countries have acquired microcomputers for analytical work (as well as for other routine office tasks).

This interest is reflected by a number of recent international meetings devoted to the topic: energy planning was one of the three areas addressed by the 1st International Symposium on Microcomputers for Development held in Colombo in November 1984 (1); the use of microcomputers for energy planning was the subject of a United Nations Department of Technical Co-operation for Development Workshop in New York in September 1985 (2); and the 1985 New Delhi meeting of the International Association of Energy Economists (IAEE) devoted several sessions to this topic. Sessions devoted to energy planning techniques have become routine at professional meetings of energy economists, planners, and modellers.

However, despite the great interest in the topic and the proliferation of "energy models" that have appeared over the past few years, it is unclear to what extent the analytical frameworks and modelling tools are in fact useful, whether from the country perspective (e.g., in terms of the contribution made to the definition of national energy strategies), or from the perspective of financial and assistance institutions. Even despite the generally poor lack of documentation, it seems clear from what is available, as well as from the experience of the UNDP/World Bank energy assessments, that the scope, appropriateness, and sophistication of microcomputer energy models vary quite considerably, and that the recent experience with microcomputer-based energy planning is not uniformly good.

Perhaps part of the reason for the proliferation of different approaches and models is the lack of guidance from the assistance community. This is in sharp contrast to many other analytical areas: for example, the early definition by the World Bank of what constitutes a minimal acceptable analytical framework for macro-economic work has been of invaluable service. The Minimum Standards Model and its subsequent revisions (3) provided clear guidance to the sort of overall macro-economic modelling efforts that were minimally desirable. This is true even though the basic version of the model could not be applied in many places without some considerable modification.

The experience with electric sector planning, particularly with respect to the WASP model for capacity expansion planning (developed and made available for many years by the International Atomic Energy Agency (IAEA)) is similar. This model is in use in many developing countries because it is well documented and because financial institutions (including the World Bank) accept it as a basis for investment decisions. Indeed, the Energy Department of the bank has recently conducted a detailed review of models used for electric power system planning (4), which provides a great deal of guidance to borrowers as to what constitutes an acceptable modelling basis for project lending decisions.

The absence of guidance on energy modelling from some of the bilateral assistance agencies that have provided long-term technical assistance to developing country energy planning agencies is particularly surprising given the substantial sums that have been spent on model development. One might note that appropriate guidance on this subject is less a question of recommending some particular piece of software as it is guidance on what constitutes an appropriate analytical process.[2]

---

[2]As part of this study, only one published source was identified that could reasonably be interpeted as authoritative guidance on computer modelling of energy systems from a source in the development banks: in a paper to the International Conference on Energy Planning in Bangladesh in 1985, King presented an excellent, albeit brief, discussion on energy modelling (K. King, "Modelling the Energy Sector in Developing Countries," International Conference on Energy Planning in Bangladesh, November 1985). Since this is based on the Asian Development Bank's (ADB) considerable experience in supporting energy planning efforts in a number of Southeast Asian countries, it contains a wealth of useful lessons.

We are not aware of any systematic review of computer models for energy planning that goes beyond a mere compilation. A 1981 report to the European Commission, for example, which sought to review models for energy planning "suitable for developing countries" (5), is largely a compendium of author-supplied descriptions of mainframe models (of which most were either for detailed electric sector planning, or large-scale models developed in the United States and Europe that were not in fact appropriate for and have never been used by developing countries).

OBJECTIVE

This section addresses the current lack of guidance on the subject of microcomputer-based energy planning. It complements the earlier proceedings of the UNDTCD Workshop on Microcomputers in Energy Planning by providing more explicit guidance on the factors that will determine the success of microcomputer-based activity. We therefore provide explicit discussion of the criteria by which models should be designed and selected, critical reviews rather than laudatory presentations of available software, information on current costs, actual examples, or hardware procurement, and frank discussion of lessons to be drawn from the case studies.

26

The Impact of Computers on Energy Planning

BACKGROUND

Computers have been an integral part of national energy planning since the mid-1970s when governments instituted such activities on a major scale. Large-scale computer models were a central part of the policy analysis and planning functions of the United States Department of Energy (and its predecessor, the Energy Research and Development Administration), and were also used extensively from an early date in the major European countries and the EEC.

Since energy planning activities began before the advent of microcomputers, it was natural for mainframe models to be introduced in developing countries. Because microcomputer use since 1981 was shaped largely by this earlier experience with mainframe models, it is useful to begin with a brief review of the evolution of such models in the United States and Europe, and with their subsequent introduction into developing countries.

REFERENCE ENERGY SYSTEM MODELS

The Reference Energy System (RES) is the conceptual basis for many energy models, both mainframe and micro. Developed in the early 1970s at Brookhaven National Laboratory as a framework for energy analysis in the United States, its key ingedient is the energy system in network form in which the production, transformation, and utilization of energy is portrayed as network links. Great emphasis was placed on the structure of end-use demand, with energy demands projected as much as possible on physical outputs (tons of steel produced, vehicle miles travelled, etc.). The network simulation version was known as ESNS (energy systems network simulator), whereas the linear programming version was known as BESOM (Brookhaven Energy System Optimization Model).

Subsequently, the model was linked with the University of Illinois input/output model, and the Hudson-Jorgenson model of the U.S. economy (6). In this form it was used for a number of years by the U.S. Energy Research and Development Administration (ERDA) as a basis for their annual planning exercises.[1]

Conceptually similar is the ensemble of models developed for the Commission of European Communities that also draws on the work at the Institut Economique et Juridique de L'energie at the University of Grenoble.[2] Well-known models such as MEDEE and EFOM have been adapted by Systems Europe S.A., a major Belgian consulting firm, into a variety of medium and long-term, national and multinational models used by the EEC and other Euorpean agencies (7).

The first major energy modelling effort occurred some years prior to this in the early 1970s, when Manne and Goreux at the World Bank Development Research Center developed ENERGETICOS, a linear program for Mexico. This was probably the first such linear programming model intended expressly for a developing country's energy sector (8).

These models profoundly influenced the subsequent development of energy models for developing countries. For example, BESOM was the conceptual basis for the MARKAL model developed jointly by BNL and the Kernforschungsanalage Julich in the Federal Republic of Germany for the International Energy Agency (IEA) (9). The MARKAL model, a very large scale, dynamic linear programming model, has since been applied to a number of developing countries. For example, the Indonesian Atomic Energy Agency has recently acquired the model for use in its long-range planning studies. The EEC models were used as the basis of the Energy Master Plan models of Thailand (10) and the BNL models were used as a basis for the Tunisia model (11). By the late 1970s, such linear programming models were very much in vogue. In the last section of Part IV, several linear programming models of this genre are reviewed.

Reference energy systems themselves, in their graphic network representation, have been widely used in developing countries, especially to portray historical energy balances. Figure 26.1 shows such an RES.

THE BROOKHAVEN MODELS

The first adaptations of the Brookhaven models to developing countries were further developments of the ESNS simulation model that exploited the network structure of the reference energy system. The LDC-ESNS model was developed for use in the Egypt and Peru energy assessments, the first two major LDC energy assessments conducted by the U.S. Department of Energy (12).

With the benefit of both hindsight and the intervening microcomputer revolution, it is easy to see why these models were not a great success. The rigid network

---

[1]See, e.g., K. Hoffman and D. Jorgenson, *Bell Journal of Economics*, 8, no. 2, 1977.

[2]For a good description of the early work in France, see, e.g., B. Laponche, "Previsions et Preparation aux Decisions en Mateire Energetique," Commissariat a l'Energie Atomique Department des Programmes, March 1978. For some of the more recent work, see, e.g., P. Criqui, M. Quidoz, and I. Hajjar, "Construction du modele SIBELIN," CRNS-IEJE, January 1985.

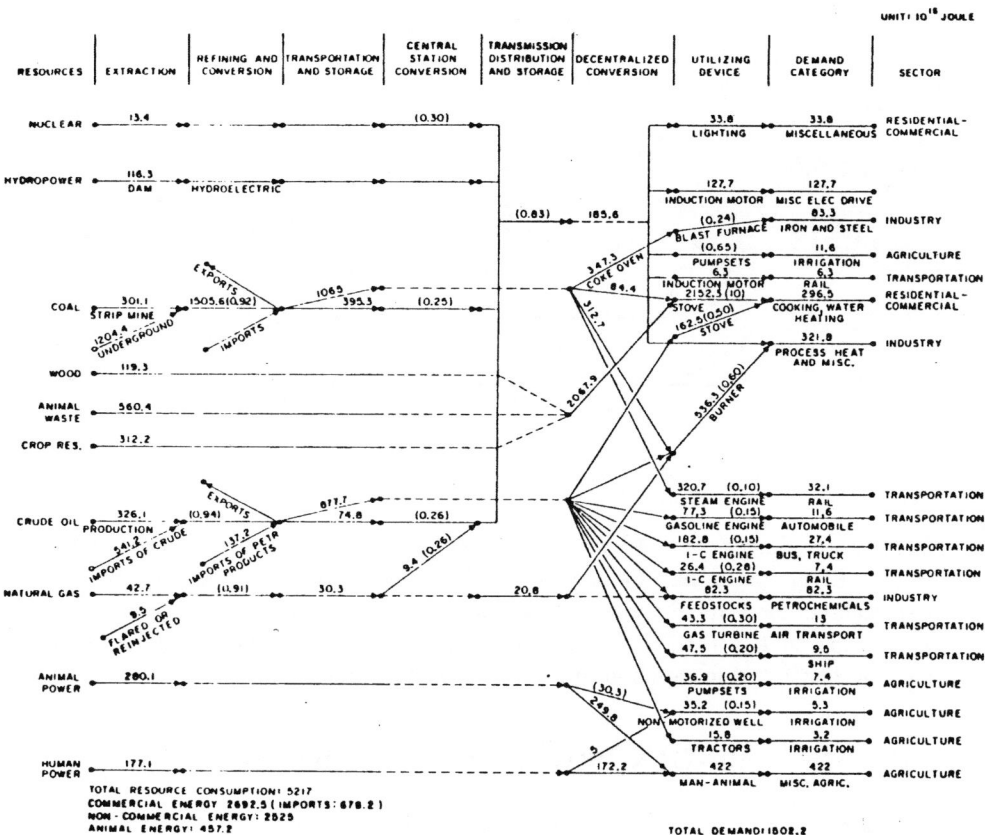

Figure 26.1 A reference energy system.

structure, in which all nodes were predefined in a cumbersome link/node numbering scheme (see Figure 26.2), lacked any kind of flexibility. And as large mainframe codes, the potential number of countries that could effectively use them was very limited. (As noted later, most energy planning entities in developing countries simply did not have sufficient access to large mainframes for the acquisition of such models to have been worthwhile).

Under the sponsorship of the Saudi Arabian Al'Dirriyah Foundation, a new version of ESNS was developed at the State University of New York at Stony Brook for use on PRIME minicomputers (under the name EDIS). Although still bound by a rigid, predefined network structure, the system made good use of color graphics and user-friendly menu techniques, and was successfully implemented in the Dominican Republic under a USAID-sponsored technical assistance project (13).

Although by the late 1970s minicomputers cost an order of magnitude less than mainframes, few national energy planning bodies could afford the $100,000 price tag (to say nothing of the associated maintenance fees). Indeed, by 1982 it was clear

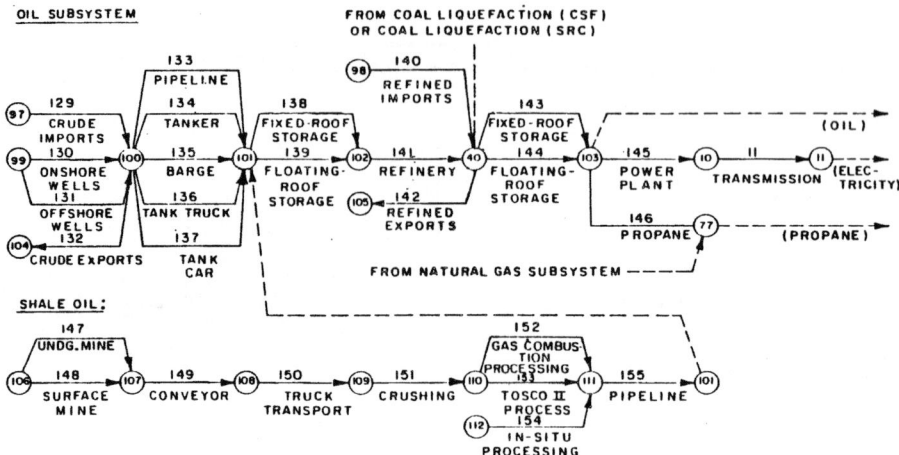

Figure 26.2 The network structure of LDC-ESNS.

that microcomputers would reduce hardware costs by another order of magnitude, and the EDIS system has not been adopted elsewhere. (In fact, five years after the installation of the PRIME computer, a maintenance agreement has not yet been signed. This is the fault of PRIME itself, which could not agree on which regional maintenance center had jurisdiction in the Dominican Republic).

LINEAR PROGRAMMING MODELS

As noted above, linear programming models of the energy system have been written for many developing countries. Table 26.1 lists a representative sample of these. Such linear programming models in fact reflect a remarkable degree of diversity, ranging from very simple, static models of between 50 and a few hundred constraints and variables (Philippines, Pakistan) to the vast, multiperiod dynamic optimization of MARKAL.

The model developed for the Thailand energy master plan is a representative example of such linear programming models. Although it is more accurate to speak of an ensemble of models that also included simulation versions and a large data base. However, these models are typical of the efforts on the part of expatriate consultants to transfer the modelling philosophy developed in the 1970s in the United States and Europe to developing countries. The models are an adaptation of the EEC models assembled by Systems Europe SA (14), whose core is a set of 15 energy networks. These networks bear a great deal of resemblance to those of LDC-ESNS (see Figure 26.2).

These networks can be run in either a simulation or optimization mode. The linear program selects that combination of resources, fuels, and processes that minimizes the total discounted cost over the period 1981–2001, subdivided into five-year intervals reflecting the fifth–eighth national development plans.

Table 26.1
Linear Programming Models

Country	Model name, author	Comments
Mexico	ENERGETICOS, Manne+Goreux for the World Bank*	
Pakistan	Riaz[†]	
Tunisia	Gordian Associates (U.K.) for the State Oil Company ETAP[‡]	BESOM derivitive with good refinery detail.
Thailand	Systems Europe[§]	See text.
Israel	OMER, Haifa University[‖]	
PRC	Energy System Model, Tsinghua University[#]	
Indonesia	KFA Julich for the Indonesian Atomic Energy Agency	MARKAL model.

*F. Garza and A. Manne, "ENERGETICOS: A Process Model of the Energy Secotrs," in Goreaux and Manne, eds: Multi-level Planning: Case Studies in Mexico: North-Holland, Ansterdaum 1973.

[†]T. Riaz, "Long Range Energy Options: Some Policy Implications for the Pakistan Economy," *Economics of Planning*, 16, no. 1 (1980), 33– .

[‡]Gordian Associates, "A Linear Programming Model of the Tunisia Energy System," report to Enterprise Tunisicnne d'Activites Petrolicres (ETAP), April 1981.

[§]Robert R. Nathan et al., "Thailand Energy Master Plan Project," report to NEA/ADB/UNDP, Feb. 1982.

[‖]A. Breiner and R. Karni, "Energy and the Israel Econony," *Technion*, Haifa, July 1981.

[#]"A Methodology for Energy Supply and Demand Evaluation," Tsinghua University, Beijung, Feb. 1984.

In fact, the EEC/Systems Europe models have a much more rational structure for potential application to developing countries than the BNL/BESOM models. The BNL models had elaborate detail for end-use devices, the nuclear fuel cycle, and the electric sector, yet refined petroleum products were characterized by a single activity "oil products," with a correspondingly primitive representation of refineries (rationalized on grounds that this subsector was the domain of the private sector whose details were of no great concern to the government). Whereas this may well have reflected the priorities of the U.S. Energy Research and Development Administration in the mid 1970s, in its application to developing countries such a balance of emphasis is clearly inappropriate.

The Systems Europe models provided a much better balance in this regard. Indeed, the optimization model is best described as a refinery optimization model to which other sectors have been added. However, Thailand's four refineries are not individually modelled, but rather aggregated into a single entity.

On the electric sector side, the Systems Europe model adopts the fairly standard LP approach. Capacity is dispatched into a four-block representation of the annual load duration curve (peak and offpeak for summer and winter).

The major limitation of the Thailand model appears to be its sheer complexity and, as least as originally implemented, its reliance on a mainframe machine at some distance from the National Energy Administration. Moreover, the overall structure of the networks is by "sector" (oil, gas, coal, hydro, and geothermal power generation, utilizing sectors, etc.), rather than by institution (e.g., in the case of electricity making, an explicit distinction between EGAT and the distribution companies, or in the refinery sector differentiating among the individual refineries whose ownership structure is quite different). To be sure, whether such institutional detail could be incorporated into an LP is arguable, but certainly for simulation purposes an institutionally oriented structure seems warranted.

LESSONS OF THE MAINFRAME ERA

In general, it can be stated that energy planning models based on mainframes have not proven very useful to energy planning agencies in developing countries, an experience that has several explanations. In some countries such models were developed at great expense, yet never used at all.

A first problem related to the "black box" nature of the models. The client was neither part of the model development process nor of the code development, with the result that the expatriate (and usually absent) consultant was the only individual who had sufficient knowledge of the code to make changes, or indeed to provide decision makers with a clear understanding of the assumptions and limitations of the model.

A second problem related to training. Many computer modelling projects consumed vast resources for computer programming, model formulation, and debugging, leaving little time for on-the-job training. Typically, models might be ready only toward the end of multiyear projects.

The third problem concerned the mainframe machines, themselves. Energy planning agencies were required to install their energy models on machines owned by others, typically, the electric utility or a national computing center. Even aside from the problems of being able to get access to such machines during normal office hours (with administrative uses almost always being given higher priority), there remained the issue of location. Since energy agencies are rarely located in the same building as those who own such machines, the incentive to use the models remained very low. To be sure, in a few rare cases remote terminals were available, but obtaining paper copy of outputs usually required a trip to another building.

As always, there are some notable exceptions to the general experience. In South Korea, the DFI model (a large-scale general equilibrium model originally developed for the U.S. Department of Energy) has been successfully implemented and modified by a number of government institutes. The refinements in the current South Korea version were made without expatriate assistance. However, this is a country where there is now a large, well-established cadre of professionals trained at major universities in economics, operations research, and systems analysis.

Such a cadre of professionals, however, is often absent in the lower income countries of Africa and the Caribbean. Thus, what is reasonable and feasible in places such as India, Argentina, Taiwan, Brazil, and Mexico may not be possible elsewhere. It can be asserted that the successful application of mainframe models has been limited to this group of more advanced countries.

THE IMPACT OF MICROCOMPUTERS

Microcomputers have been introduced into energy planning over the past five years at a rate that has matched its introduction into most other fields such as health, finance, business, and agriculture. Today microcomputers are a part of energy planning in almost all developing countries.

The reasons for the acquisition of microcomputers on such a near-universal scale evolve in large part directly from the shortcomings of the earlier mainframes. Mainframes were too expensive for energy planning agencies. Microcomputers are very cheap (especially when compared to the labor costs of technical assistance). Mainframes were in other buildings at other institutions. Microcomputers sat on the desks of energy planning agency staff. Even if mainframe models were available, most routine calculations still needed to be done by hand with only calculators. Microcomputers enabled routine calculations (and other tasks such as word processing) to be done more efficiently.

In fact, a number of reasons make microcomputers ideally suited to energy planning. First (as noted earlier) is the issue of data. The type and quantity of data useful for energy planning are well suited to the capacity of low-cost storage media associated with the current generation of microcomputers: 20–30 megabyte hard disks that cost about $300–500. Moreover, microcomputer data base software such as LOTUS and dBase prove to be very capable of manipulating typical energy and economic data sets. Energy planning does not require data bases in the gigabyte size range or the much more sophisticated data base software typical of mainframes that are needed to manage such data bases.

The second reason concerns the natural match between many typical energy analysis tasks and a spreadsheet-based programming environment. Demand forecasting, macro-economic modelling, project analysis, and the preparation of energy balances are all functions ideally suited to spreadsheet calculation. To be sure, there is no reason why spreadsheets cannot be run on mainframes, but sophisticated spreadsheets uniquely matched to the capabilities of microcomputers and their users.

In a similar vein, even though it is true that mainframe computers could be as user-friendly as modern micros and offer the user similar (if not superior) graphics capabilities, the fact remains that from the very start microcomputer software was developed with the layperson user in mind rather than highly trained programmers or engineers. The mass market for microcomputers (and hence their low cost) would not have been possible without user-friendly software designed for general purpose application. This means that the type of staff typical of developing country energy planning agencies, who often lack advanced degrees in computer-related fields, are perfectly capable of learning how to use both the hardware and software within some minimal time period. This enhanced ability to train individuals to use microcomputers is an impact of a much more fundamental nature than the mere low cost, accessibility, and desk-top scale.

Microcomputers have reversed the traditional decision making and resource allocation priorities associated with computing. In the mainframe era, the order of priorities was: hardware aquisition, software, then training.

This order corresponded to the relative costs. Today the order of priorities, and cost, has been reversed: training, software, then hardware.

As we see later in the case studies, even the hardware outlays have come to be viewed as operating costs, not capital expenditures. The lower priority given to hardware choices simply reflects the circumstance that even a major microcomputer purchasing error can be corrected at relatively low cost.

Another quite fundamental impact of microcomputers has been the associated developments in software. Even in the case of the minicomputers of the late 1970s, which typically sold for \$100,000–\$200,000, software was expensive, and in most cases needed to be custom designed, involving expensive computer programming time. For example, a FORTRAN 77 compiler for the PRIME minicomputer was priced at about \$7,000. Yet today similar FORTRAN compilers for microcomputers can be aquired for \$200–\$300. Even more important, one can purchase at comparable prices ready commercial software for most analytical tasks, ranging from linear programming and project analysis to presentation quality color graphics. Thus microcomputers enable an energy planning agency to perform a range of analytical functions by simply buying low-cost commercial software that in the mainframe era was not possible because of cost.

Thus the initiation of microcomputer-based energy planning and its success is largely a matter of the design of training programs, since the hardware and software costs themselves have become increasingly small. As we see later in the Morocco case study, microcomputer training has become the focal point for integrating the activities of the subsectoral agencies and the energy planning agency itself.

MICROCOMPUTERS AND DECISIONMAKING

If microcomputers have made easier the task of energy analysis and extended its range, to what extent has the quality of decisions actually improved? For all of the much advertised emphasis on presentation quality graphics, to what extent has the communication between analyst and decision maker improved? Indeed, to what extent have decision makers themselves used microcomputers? These are the most important questions to answer because if the only impact of microcomputers has been to do analysis quicker, or do more analysis within the same resource limitations, then microcomputers will not have proven to be very important.

Unfortunately, these are also the hardest questions to answer. In part this is because the widespread use of microcomputers is so recent that it is hard to come to firm conclusions at the present time. It is certainly true that to date senior decision makers have made relatively little use of microcomputers for actual decision making. To be sure, heads of planning agencies, and perhaps even one or two ministers, may have a microcomputer at his or her desk, but they are likely to be used to retrieve data and ease writing tasks. In fact almost everywhere decisions are still made on the basis of technical analyses prepared by technical staff. Early expectations that decision makers would themselves run models have not been fulfilled. However, one might note that this is not unique to developing country energy planners; most chief executive officers of major U.S. corporations do not make decisions on the basis of their own microcomputer simulations either.

27

Criteria for Developing and Selecting Energy Models

One of the main difficulties faced by national energy planning agencies in implementing a microcomputer-based planning approach is evaluating the mass of literature advocating particular software packages and models. Sales pitches for "better" models by groups and consultants from the developed countries can frequently be intimidating. The objective of this chapter is to discuss the issues that are important to the design, implementation, and selection of microcomputer models for energy planning, and our review of some selected software packages presented in the next chapter uses these criteria as a guide.

Table 27.1 presents a summary of the most important issues to be addressed in an energy modelling activity. The need for a discussion of such questions is evident from the shortcomings of much of the modelling work that has been done over the last few years. For example, it is astonishing that several energy planning software packages do not produce, as one of the standard outputs, an energy balance in the UN/OECD/Bank format. Each of the criteria presented in Table 27.1 is discussed in the following paragraphs.

INTEGRATION WITH THE PLANNING PROCESS

Energy planning in developing countries has several distinctive characteristics that are central to the definition of a modelling framework. The first is the hierarchial nature of planning, in which at least three levels are of interest here.

1. Macro-economic planning, which integrates planning at the sectoral level to produce a consistent macro-economic picture.
2. Energy planning, which integrates energy subsector planning (electricity, petroleum, fuelwood, etc.) into a consistent energy sectorwide picture.
3. Subsectoral planning, conducted by the implementing agencies (electric utilities, refineries, and petroleum planning).

Table 27.1
Summary of Evaluation Criteria

General Criteria

Integration with the Planning Process

What kinds of uncertainty can be evaluated exogenous (e.g., world oil price), climatic (e.g., can one assess the impact of droughts), policy (e.g., uncertainty about the demand response to tariff reforms), geological (resource uncertainty)?

Macro-economic Model

Integration with Official Government Forecasts

Can the model be aligned with the official macro-economic forecast of the government? Does it follow the same sectoral disaggregation? How does the presentation of the external accounts mathc that of the central bank?

Disaggregation

How is the disaggregation between energy and nonenergy sectors achieved, and how is it made consistent with the official government forecast that may not nicely distinguish the "energy" sector?

Alignment of Model to Data

In the more sophisticated macro-economic models, to what extent is the model formulation consistent with data availability? For example, in production functions, how are the substitution elasticities derived?

Energy Model

Project Level Information

Does the model incorporate project level information? If yes, how are the projects assembled into portfolios (i.e., how are alternative investment options defined?). Can the model accept such plans directly from the subsectoral institutions?

Demand Structures

How flexible is the demand structure? How do major fuel substitution and/or conservation projects affect the demand structure? Limited to aggregate econometric equations or can end-use detail, where available, be integrated?

Technical Representation of the Electric Sector

Does the model have a sufficient level of technical detail to be credible to the electric utility: e.g., does it include a formal optimal dispatch algorithm in which plants are dispatched into a load-duration curve according to merit order, and according to plant availabilities set forth by the capacity expansion plan, or is it limited to simple "generation mix" or econometric estimations of fuel consumption'?

Output Format and Conventions

Is the output in a standard and conventional format (e.g., does the model produce an energy balance table in the standard UN/OECD format?) Does it use the standard conventions for energy data (e.g., as set forth in the World Bank's Energy Department Paper on Energy Data?)

Institutional Structure

Does the model accurately reflect the institutional structure (e.g., can the model distinguish between a gradually growing, interconnected electric utility grid, and dispersed systems? Can the model distinguish between multinational and national owned oil companies?). Does the model incorporate a sufficient level of detail for pricing studies (e.g., where quite different price structure exist for premium, regular, and aviation gasoline, is the model capable of this disaggregation?)

Petroleum Sector Representation

How does the energy model determine petroleum supply mix? In countries with more than one refinery, is each treated separately? Does the model distinguish among crude types, and how is the refinery output determined (aggregate "yield" coefficients, or more explicit representation of individual unit operations)? In the case of mismatch of production and product demands, can the model handle spiked and/or reconstituted crudes?

Table 27.1 (*continued*)
Summary of Evaluation Criteria

Noncommercial Fuels

How are noncommercial fuels treated in the energy model? How does the model accommodate substitutions between fuelwood and petroleum products?

Financial Detail

Beyond the overall impact of consumer price level as a determinant of energy demand, to what extent are the financial flows associated with particular tariffication structures captured in the model? How does the model reflect such features as cross-subsidization mechanisms on petroleum products, royalty/production sharing arrangements for potential fossil resource development programs, and the overall flow of funds from the energy sector to and from the government?

Subsectoral Models

Data Integration

To what extent does the model accept data from models at the subsectoral level (fuelwood models, refinery optimization models, etc.)?

The second characteristic concerns the nature of the energy sector planning function itself, which is primarily one of harmonizing the detailed plans of the implementing agencies with overall macro-economic plans. Energy planning conducted in a vacuum without close coordination with these other bodies is not likely to remain anything but an academic exercise, no matter how rigorous and sophisticated its models are.

It follows that the modelling framework should also be consistent with this planning process. Thus a modular, hierarchical modelling system in which individual models can be exercised independently, as well as in concert with each other, is to be preferred over a large, single, all-inclusive model. Ideally, the models at the third subsectoral level should be run jointly by both the subsectoral institution and the energy planning institution, or at the very least the energy model must have the ability to accept information at the same level of detail as that offered by the subsectoral entity. For example, in the case of the electric sector, the model should have the ability to accept the capacity expansion plan of the electric utility.

TREATMENT OF UNCERTAINTIES

Energy and macro-economic planning in the typical oil-importing country is beset with a multitude of uncertainties, and the ability to identify energy and investment strategies that are robust under a variety of future outcomes is perhaps the key element in a successful planning process. Several fundamentally different types of uncertainty must be faced: (1) the stochastic character of all natural phenomena (such as rainfall and streamflow, or the distribution of oil reservoir sizes), (2) the uncertainties of the international economic and energy environment over which the decision makers in typical developing countries have no (or at best only marginal) control, (3) the uncertainties in the functioning of the domestic economic system (and hence uncertainty in the responsiveness of the system to policy initiatives).[1]

There is a well-established body of analytical techniques that deals with the first type of uncertainty, usually referred to as "decision analysis"; and indeed these techniques are widely used in subsectoral energy models for petroleum exploration, probabilistic simulation of electric generating systems, and the management of multipurpose water, resource systems. The successful application of such techniques depends on the ability to define the underlying probability distributions, which in turn demands an adequate data base. Despite the usual problems of data in developing countries, however, experience shows it to be much easier to extrapolate probability distributions for natural systems (from other countries or similar climate and geologic regimes) than to extrapolate socioeconomic phenomena. Thus these techniques have been applied fairly successfully to developing countries.[2]

Application of such techniques to the features of the international economic system are much more difficult because the laws of physics do not apply to political, economic, and social behavior; hence a derivation of the necessary probability distributions becomes an almost impossible task. For example, all kinds of game theory-based models were developed in the late 1970s to model OPEC behavior and the future trajectory of world oil prices, yet few, if any, came even close to projecting the oil price downturn in the 1981–1983 period, much less the sudden collapse that followed later. Such models have not been used anywhere in developing countries as a basis for decision making.

In response to these difficulties, the standard technique has been the use of scenario simulation in which values of such exogenous factors are hypothesized without any initial judgment on the likelihood of occurrence. In effect, the judgment about uncertainty is shifted from the analyst to the decision maker, since it is the latter who must make the tradeoffs between scenario and policy response on the basis of the projected impact.

Such an approach, however, if it is to be practical, poses some rather severe limitations on a prospective modelling system. First, since there may be several exogenous scenarios for each of several parameters, the possible permutations of scenarios quickly proliferates. The analyst must be able to quickly test out a multitude of scenarios, discard those with duplicative or uninteresting outcomes, and focus on a more manageable subset. Too many scenarios will simply overwhelm the typical decision-making body. If such a model is to be used in practice, a simple rule of

---

[1]For a further discussion of the treatment of uncertainties in energy planning, see, e.g., P. Meier, *Energy Planning for Development Countries: An Introduction to Quantitative Methods.* Boulder CO: Westview Press, 1986.

[2]Thus, multinational oil companies engaged in exploration and development efforts in developing countires can use exactly the same set of analytical tools to evaluate seismic or aeromagnetic survey data as they do in Texas. The set of tools suitable for analysis of, say, electricity demands in a developing country, however, are quite different from those appropriate for Texas.

thumb would be the requirement that a complete model run consume no more than 20 minutes.[3]

INTEGRATION WITH MACROECONOMIC PLANS

At the macro-economic level, perhaps the most important attribute is the ability to replicate the official government forecast. In most countries there exists an official plan, typically prepared in an annual or five-year cycle by a ministry of planning. Whereas such forecasts may well be in the nature of a set of goals rather than a forecast and therefore subject to some controversy, it is nevertheless true that as a matter of practice this forecast usually has official standing, and therefore provides the basis for the participation of the ministry of planning (or whatever may be its particular name) in sectoral planning efforts.

It follows that the macro-economic model used for the short to medium term covered by this forecast should have the ability to replicate almost exactly the official forecast as the baseline for analysis. Ideally, this baseline should then be capable of perturbation to reflect alternative assumptions in a consistent fashion. Moreover, the presentation should be in a format that closely follows the official presentation (e.g., using the same sectoral disaggregation, even if this poses some element of difficulty further down in the hierarchy of models, such as for the demand projections).

DISAGGREGATION OF ENERGY AND NONENERGY SECTORS

There are two basic approaches to the reconciliation of highly aggregated macro-economic models (in which energy and nonenergy sectors are either not distinguished at all, or at a level of disaggregation unsuited to energy work) and energy models. The first approach is to run macro-economic and energy models independently, with energy demands driven by GDP assumptions, but without a formal link from the energy model to the macro model. In such a scheme, the energy model impacts (debt service, investment, etc.) are simply compared to the aggregate projections from the macro model without concern for consistency (see Figure 27.1, Scheme A).

A more detailed but also more difficult approach is to attempt some explicit disaggregation of the macro model. Where formal models are concerned, this may re-

---

[3]Obviously, what matters here is the effective turnaround time, not the CPU seconds on a mainframe actually required for the computation. A related issue concerns the tendency to build ever larger models as the capability of hardware increases, and execution time per instruction decreases. In the mainframe energy model era of the 1970s, models grew ever larger as super computers and highly efficient LP solvers (such as the CDC APEX series) reduced computation times by orders of magnitude: print outs for overnight runs required handcarts to bring to the analyst's office in the morning. Another repetition of this cycle seems imminent, as desk-top microcomputers exploit ever more powerful microprocesser designs.

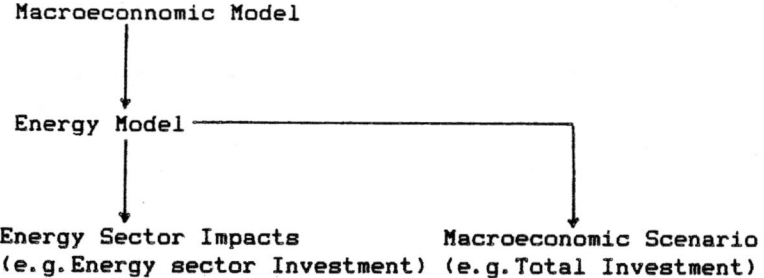

Scheme A:

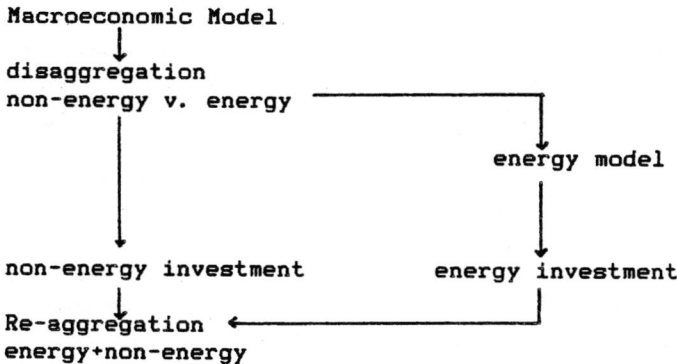

Scheme B:

Figure 27.1 Energy-macroeconomic linkages.

quire extensive changes in the equation structure. On the other hand, where the basis is a macro-economic accounting framework rooted in the official government plan, it is usually quite easy to subtract energy sector investment (and debt service) from the aggregate investment projections. Moreover, since the energy sector often contributes a negligible proportion to GDP, its inclusion in other valued-added sectors poses few problems and an explicit disaggregation here is not necessary. (Electricity is usually included in a sector such as "municipal services," which also includes water and wastewater utilities; the refinery is frequently included in the industrial sector).[4] This approach is illustrated on Figure 27.1, Scheme B).

---

[4]In Sri Lanka, for example, the entire sector "electricity, gas, water, and sanitary services" accounted for only some 1.3% to total value added in the economy. This is, of course, in sharp contrast to the corresponding importance of this sector to investment and debt service.

ALIGNMENT OF THE MODEL TO AVAILABLE DATA

Any macro-economic models developed for use in energy economic studies prove to be poorly designed from the standpoint of possessing a level of complexity that is consonant with the available data. Even those models that make no claim of relevance for the short to medium term, but whose focus is the long-run structural adjustment process, frequently use theoretically convenient equations whose coefficients are essentially impossible to derive from actual, in-country data. Such models typically use log-log production functions involving constant substitution elasticities (among energy, labor, capital, and "other" material inputs) that are quite far from the sort of production functions estimated in sectoral studies.

This problem is perhaps most acute in that subset of countries whose economies are still largely agricultural, and whose export earnings depend heavily on a few export crops. Any reasonable and useful formulation of plantation sectors involving perennial crops would need to include in a sectoral production function, beyond price and energy (and/or fertilizer) inputs, such variables as the area under cultivation and the age distribution of the planting stock. Since in many cases increases in the area under agricultural or plantation cultivation implies a decrease in the natural forest area, there may also exist an indirect link to the fuelwood supply. In some cases (such as rubber), the replanting cycle itself contributes to the fuelwood supply. (In Sri Lanka, for example, some 70% of the urban fuelwood supply in the Colombo area is rubber wood, i.e., trees felled to make way for new plantings).

As an illustration, consider the production functions of the so-called Cobb-Douglas type, which are frequently encountered in such macro-economic models because they are the simplest forms that allow factor substitution:

$$x = a \, L^b \cdot E^c \cdot K^d$$

where x is incremental output, L is the incremental labor input, E is the incremental energy input, and K is the previous year(s) investment, and a, b, c, and d are technical coefficients. Yet production functions encountered in typical sectoral studies are frequently of a quite different form, especially for agricultural sectors, including, at a minimum variables for land area, fertilizer inputs (of importance anyway to energy planning because of the energy intensiveness of their manufacture), and irrigation/rainfall inputs.[5]

The argument here is not that macro-economic models should necessarily be encumbered with vast sectoral detail. However, to the extent that these models are built to examine the longer term issues of structural adjustment (and the impact of

---

[5]For example, R. Bandaranaike, "Tea Production in Sri Lanka: Future Outlook and Mechanisms for Improving Sectoral Peformance," Central Bank of Ceylon Occasional Papers #7, 1984. The production function takes a log-linear form that includes variables for land area, fertilizer application, rainfall, raindays, and a series of dummy variables for elevation. Similarly, the model for the rubber sector in M. Hartley, M. Nerlove, and K. Peters, "The Supply Response for Rubber in Sri Lanka: A Preliminary Analysis," World Bank Staff Working Papers #657, World Bank, 1984, included regression variables that captured the age of the planting stock and the extent of subsidies.

that adjustment on the energy sector), maximum use should be made of the sectoral models built by others interested in the rehabilitation or adjustment of major producing or exporting sectors.

PROJECT LEVEL INFORMATION

Project definition and energy sector investment planning is one of the primary functions of a national energy planning entity, yet, as noted in the UNDP/World Bank energy assessments, energy sector investment planning is frequently one of the weakest areas in energy planning. It follows that one of the most important questions concerning the suitability of a general modelling framework is the ability to include project level information. The smaller the country in question, the more important does this issue become, since a single electric sector project may, in such cases, account for a major share of public sector investment and debt service.

It should be noted that this need for project scale analysis and investment planning is not met by the simple availability of a project analysis package. There are surely dozens of commercial packages now available for IRR calculations and the like. Equally important are questions such as how alternative investment plans are assembled, and once assembled, how they are integrated into the overall analysis. Are project portfolios assembled on an ad hoc basis, by ranking of IRRs (in fact, a possibly erroneous procedure if capital is constrained), or by some sophisticated capital budgeting model (maximize net present value subject to capital constraints)?

Especially important are such questions for the electric sector, where the determination of the optimal capacity expansion path is a matter of considerable complexity. The main issue here is the degree to which the energy model can be integrated with typical electric utility capacity expansion plans.

DEMAND STRUCTURES

The adequacy of the demand structures in many models is in some doubt, particularly with respect to the ability to estimate the impact of policy initiatives. Perhaps the best example concerns the electric sector, which in many developing countries is supply constrained. The major determinant of demand is the ability of the electric utility to extend the system into both previously unconnected rural areas and rapidly expanding cities. Yet the traditional econometric formulation of demand as a function of price and income frequently ignores this issue (quite aside from such other assumptions as constant own-price elasticities, and usually absent cross-price elasticities).

To be useful for an energy planning and policy exercise, the demand structure needs to consider, beyond just price and income variables, such factors as the extent of energy and peak load accounted for by auto-generation (which is highly sensitive to the level of reliability provided by the central system), the theft rate (as distinct from purely technical losses, which are subject to altogether different types of policy intervention), and the rates of new connections (in turn also related to the considerable costs of distribution system expansion).

TECHNICAL REPRESENTATION OF THE ELECTRIC SECTOR

Because of the importance of the electric sector from the standpoint of investment and debt service and the inherent technical complexity of the operation of this subsector, an important issue concerns the degree to which an energy model has sufficient technical credibility to demonstrate the impact of multiple and simultaneous policy and project initiatives. It is not at all uncommon for a development bank, together with other multi- and bilateral agencies, and the country government itself, to be contemplating the financing of additional generation capacity, the rehabilitation of the distribution system to reduce system losses, and an industrial sector energy efficiency initiative designed to improve power and load factors.

It follows that if an energy model is to be useful, it has to have the ability to demonstrate the interactive impacts of such diverse but parallel initiatives. This, in turn, requires a level of technical detail that at a minimum includes the ability to replicate the essential features of system load duration curves and the optimal dispatch of specific generation units into such a curve.

OUTPUT FORMAT AND DATA CONVENTIONS

Since one of the major reasons for using models is to increase the productivity of the staff engaged in the support work for decision making bodies, it follows that models should, as far as possible, adopt the usual reporting conventions in their presentation of key results. For example, energy balances should be generated in the form almost universally accepted by international institutions—as set forth, for example, in a recent report by the World Bank Energy Department as the standard for the UNDP/Bank energy sector assessments. This may be a somewhat obvious point, but several of the models reviewed in this section not only require manual transcription of energy balances, but in some cases require additional calculations as well.

INSTITUTIONAL STRUCTURE

Beyond the question of how the overall modelling system is aligned to the institutional realities discussed above is the degree to which an energy model itself is structured around the institutional structure. Since the revenue, tax, and subsidy flows, a major concern to government decision makers and to an economically efficient pricing system, are related to the transactions between specific energy-sector institutions, it follows that the energy model, to be useful for short to medium-term analysis, should have as its basic building blocks not subsectors (petroleum, electricity, gas, coal), but institutions (the refinery, the electric utility, the LPG company, etc.).

PETROLEUM SECTOR REPRESENTATION

The representation of the petroleum sector in energy models is frequently unsatisfactory. Problems range from inadequate disaggregation (which makes them difficult to use as a basis for pricing studies) to an inadequate treatment of refinery flexibility (petroleum product production from a domestic refinery is given in some future year by multiplication of the crude run by some set of yield coefficients). Nowhere is the ability of an energy model to interact with a more detailed subsectoral model more important than the petroleum sector. This is one of the few areas where LP-based energy models have an inherent advantage in that they can be structured around a linear programming representation of a refinery.

Simulation model representations of refineries must therefore have either an appropriate interface to a detailed refinery LP, or find some other scheme to deal with the refinery capacity expansion/flexibility issue. Treating the refinery as a black box with product outputs given as linear functions of the crude inputs is problematic even for simple hydroskimming refineries because crudes can be spiked over quite wide ranges.[6]

NONCOMMERCIAL FUELS

There are a number of alternative strategies to deal with noncommercial fuels. Many LP energy models simply ignore them. Others include them in the overall scheme for computing energy balances, but with demand projections for fuelwood and charcoal not well integrated with the commercial fuel demand projections (the issue being the substitution effects among the commercial and noncommercial fuels). Reference energy system models generally do provide a structure to maintain a consistency in such substitutions, but are ill-equipped to relate the pace of change to the price environment. And in hierarchial systems that include a detailed fuelwood model, there are special difficulties in forcing consistency between projections for electricity, petroleum products, and noncommercial fuels.

---

[6]The assumption in most simulation models is a linear relationship of the form

$$x = \Sigma \, q \cdot y$$

where q is the crude run, y is the yield coefficient, and x is the output of the i-th product. The principal difficulty is that the inputs (the q) must be exogenously defined, yet in reality the inputs will be defined as a function of the composition of the product mix.

DATA INTEGRATION WITH SUBSECTORAL MODELS

The proposition that energy models should be linked to more detailed subsectoral models cannot be argued, and the claim that a given model is consistent with other models is almost always made. Indeed, as a matter of computer technique, it is not too difficult to transfer data from one model to another. In practice, however, the question is not one of modelling but of institutional relationships. What matters is not only the technical capability of an energy model to accept subsectoral detail (as argued above in the case of the electric sector capacity expansion plan), but an institutional mechanism that ensures a sufficient dialog for the transfer to actually take place. In some countries (such as Morocco and Sri Lanka reviewed in the case studies section), such a mechanism is present; in others it is notably absent, which makes any modelling effort somewhat questionable.

RELATIONSHIP TO STAFF CAPABILITIES

Finally, in terms of application in-country, we would be delinquent if we omitted mention of the issue of staff capabilities and training. Even a modelling framework that possessed all of the above attributes would be without value unless the capability exists to use it. The working level staff must be sufficiently well trained to be able to operate it successfully. One might note that this is not only a matter of training in modelling and the particular software implementation, probably the least important aspect if the software is well designed. The central issue is the technical background and experience of the staff. A well-trained engineer or economist with some years of practice can usually be trained successfully to use a modeling system, whereas the converse is usually not true. Programmers, modellers, and others without experience in a refinery, electric or gas utility, or the like, whatever their computer expertise, rarely have the technical judgment to successfully operate a model ensemble (although the participation of a programer/system analyst may well be desirable if present as part of a broader team).

28

Software Packages for Energy Planning

BACKGROUND

With microcomputers in use throughout the world in energy planning activities, a comprehensive review of the software currently in use has become a virtually impossible task. The presentation of software in this chapter (and in Chapter 29 on software in selected subsectors) is therefore presented in the nature of a selective review. The intent is to give the prospective users of energy planning software some assistance in evaluating the strengths and limitations of alternative approaches.

It should also be noted that advantages in hardware and in the development of general purpose commercial software continue to occur at a rapid pace. Therefore energy models, too, are changing at a rapid pace with models developed as little as five years ago being replaced by new and more powerful packages that exploit the hardware advances. Thus in the few years since the 1985 UNDTCD workshop on Microcomputer Applications for Energy Planning in Developing Countries, many new models have been developed, and some of the models presented in the conference proceedings have been replaced.

THE ENVEST MODEL OF MOROCCO

The ENVEST model was built for the Ministry of Energy and Mines in Morocco by Decision Sciences, Inc., as part of a long-term energy planning technical assistance project sponsored by USAID. The ENVEST model has the distinction of being the first comprehensive microcomputer-based energy analysis system implemented in developing countries. Work began in 1980 when many were still quite skeptical about the realiability and ability to service such machines in developing country conditions.[1] ENVEST was written first in FORTRAN for the Apple II, and subse-

---

[1] In 1983 at the Reston Conference on Energy Planning in Developing Countries sponsored by USAID, many voices doubtful that microcomputers could be adequately maintained and serviced were still heard (in a panel discussion devoted to the topic of microcomputer applications). By 1984, however, there was a rather sudden change, as evidenced by the consensus that emerged at the November 1984 Symposium on Microcomputers in Development held in Colombo, Sri Lanka.

quently rewritten in Apple Pascal 1.1; a version for the IBM PC was also prepared for Costa Rica.

ENVEST's main strength is its close integration with the energy planning process in Morocco, and it was designed with the specific role assigned to the Ministry of Energy and Mines in mind. This role is focused on project analysis and the review and coordination of projects submitted by implementing agencies. Considerable attention is also paid to the impact of uncertainty in this analysis. In fact, as noted by Gorden (15), ENVEST is as much a planning approach as it is a particular piece of software.

Model Design

The most important feature of ENVEST is its integration of project analysis into the energy model. Each major energy project has its own project analysis routine (that calculates the internal rate of return). When the energy simulation is performed, there is a call to a project "portfolio" generator, in which subsets of these projects can be combined into portfolios, and whose energy and financial impacts are integrated into the overall system.

Uncertainty in the project parameters (such as capital cost) is incorporated through a Monte Carlo scheme, with internal rates of return generated as probability distributions. One problem with this treatment of uncertainty is the difficulty of specifying probability distributions: since there is no real database, the assumptions can be quite arbitrary. The basis for specifying a probability distribution for an oil shale or coal development project seems somewhat unclear, and in such cases does not appear to offer any particular advantage over a sensitivity analysis approach in which IRR is displayed in graphical form as functions of key assumptions.

Treatment of the Electric Sector

The details of the electric sector algorithm are difficult to establish because a definitive documentation has not yet been published, and because no output display is generated that clearly shows the dispatch levels for each plant. The load-duration curve is defined as a fifth order polynomial, and plants are dispatched into this curve in a predetermined merit order. Thus the plant factor is predetermined in the project analysis (since the energy output needs to be known for purposes of calculating the revenue stream from each project) and, in effect, the dispatching process occurs in terms of energy (in which the area under the curve is simply filled up by each plant, from the bottom, in merit order), rather than power levels subject to availability factors.

This procedure is sufficient only for a first-order estimate of the fuel consumption associated with a specific electricity demand in a purely thermal system. Especially in mixed hydrothermal systems such a procedure is insufficient, because only a more rigorous optimal dispatching can properly simulate hydro plant operation. In reality, it is the economic dispatch procedure that determines the aggregate annual plant factor through a consideration of the capacity availabilities in each block of the load curve. Thus the amount of energy that is, in fact, generated by a specific plant in a given year ought to be an input to the project analysis, passed from the optimal dispatch routine, not the other way around. Finally, the shape of the load-duration curve is fixed, and one cannot therefore examine the load curve (or system load factor) changes that follow from demand management strategies.

Strengths and Limitations

ENVEST meets several of our primary criteria: its close alignment to the planning process (in which the overall modelling approach is closely geared to the data that is transmitted to the Ministry of Energy and Mines), its incorporation of project-level detail, and its recognition of uncertainty. As a completely menu-driven system, it is also very user-friendly.

Unfortunately, there are a number of limitations that outweigh these strengths and limit its use elsewhere. The Morocco version in particular, being written in the now obsolete Apple Pascal 1.1, would require considerable effort to be transferred elsewhere. But much more limiting are certain constraints imposed by the programming philosophy, the most notable being that much data is hard-wired into the source code. For example, the inventory of existing power plants is contained not in data file, but in a compiled subroutine. Similarly, the program is utterly inflexible with regard to the number of fuels or the number of GDP sectors, because the arrays in question are not defined dynamically, but are again hard-wired into the compiled code.[2] To be sure, such an approach to data security certainly minimizes the possibility of user error: however, if the staff is not capable of proper maintenance of data files, then clearly a computer modelling effort is inappropriate in the first place.

A number of other problems would also limit its potential use in other developing countries. First, the output tables are quite far from the usual data format (both for energy balances as well as macro-economic information) and do not follow the usual UN/Bank format. Moreover, there is no capability to transfer output data to other software programs for reworking, either as graphs or as tables. Second, the model is not suited to (and indeed not designed for) pricing analysis. Finally, the macro-economic module, once again for reasons of being completely hard-wired into the system, is quite inflexible.

Despite these shortcomings, ENVEST should be viewed as the pioneering effort in the field, from which most valuable lessons have been drawn. Indeed, the new Morocco energy model, using LOTUS and more modern hardware, embodies much of the ENVEST approach.

---

[2]It should be noted a considerable difference exists between data entered in a spreadsheet table and data hard-wired into a compiled program. In the former case, the data is completely transparent, and any user trained in the spreadsheet software should be capable of inserting, say, an additional row or column to reflect the need for an additional fuel or sector. However, even a very experienced and expert programmer would encounter difficulties in making consistent changes to array dimensions, array indices, and labels in a very large Pascal code (ENVEST consists of some 50,000 lines of compiled code).

RESGEN

RESGEN is one of the more widely used software packages for energy planning in developing countries. In addition to Sri Lanka (an application presented in more detail in the case studies later in this report), it has been used by national energy planning agencies in Indonesia, Thailand, Haiti, the Dominican Republic, Uruguay, Pakistan, and Morocco, among others.[3] As with ENERPLAN, RESGEN is more accurately described as a software package to generate models than as a model per se. There are two versions of RESGEN in current use. The original was written in a mixture of FORTRAN and assembly language (this version is reviewed here). The newer version, written entirely in LOTUS 1-2-3 (and implemented in Morocco and Pakistan), is discussed in the Morocco in Chapter 32.

The RES network is sliced vertically into vectors (or arrays). As one moves from right to left (from demand to supply), each successive vector is calculated from its predecessor by a matrix transformation.

Special operations, such as the electric sector dispatch and the incorporation of project level information, require, of course, some special operators. And indeed in the software package itself the user does not deal at all with matrix algebra.

RESGEN is designed to be used with LOTUS 1-2-3. Input information can be drawn from LOTUS spreadsheets, and output information can be transferred to LO-TUS (or to ASCII files) for further graphics presentation. For example, all the energy sector information of relevance to the macro-economy (investment, debt service, imports and exports, etc.) is put into a LOTUS readable file for further processing in any macro model. Model verification is made easy by the fact that intermediate output can be generated at every stage of the network calculations. The supply/demand balance is presented in conventional form and can be custom designed by the user.

RESGEN is designed to be used as part of hierarchial modelling systems, and therefore requires some sort of macro-economic driver. In the case study we review the macro-economic accounting framework used for this purpose in Sri Lanka (Chapter 28). The unique feature of RESGEN is that it allows three different types of demand structure (that can be used simultaneously): (1) econometric specifications (similar to those used in ENERPLAN; see below), (2) project-specific demands that allow the user to modify demands according to the specific fuel inputs (or outputs or fuel substitutions or fuel use reductions from conservation, and (3) process models in which fuel demands are built up from a projection of end-use devices.

Treatment of the Electric Sector

The basis for the electric sector calculations is a plant-specific dispatch into a linearized load-duration curve. The dispatch algorithms dispatches individual plants into

---

[3]For a complete documentation of a RESGEN implementation, see, e.g., "The Morocco Energy Model," Ministry of Energy and Mines, Rabat, 1987.

the vertical blocks of the load-duration curve based on the respective capacity availabilities. Again with the institutional considerations in mind, the plant list takes the form of alternative capacity expansion plans as might be generated by a more detailed capacity expansion model such as WASP. There is explicit consideration of the energy constraints of hydro plants, and the algorithm can be shown to provide the optimum dispatch in mixed hydrothermal systems.

The three-block linearization proves to be sufficient for a fairly good replication of actual system dispatching. In the Dominican Republic, validation of the model against known 1985 data showed deviations in the 0–5% range at the level of specific plants. Moreover, because the shape of the load curve and the input assumptions regarding forced outage rates and the like are fully under the user's control, the model has been successfully used to examine the impacts of demand management programs, drought conditions, and system rehabilitation on electric utility fuel use and investment requirements.

Strengths and Limitations

RESGEN is a model of great flexibility, but as such it imposes rather severe requirements on the user. Familiarity with reference energy systems is highly desirable (and indeed the program user manual devotes much space to a discussion of basic energy analysis concepts). Compared, for example, to ENERPLAN, the user requires considerably more in the way of up-front training before he or she can successfully implement a model. Indeed, experience with the early versions of RESGEN pointed to the shortcomings of any compiled program and led to the development of a spreadsheet-based program (as discussed in the Morocco case study, Chapter 28).

ENERPLAN

The ENERPLAN software package was developed by the United Nations Department of Technical Cooperation for Development with the cooperation of the Tokyo Energy Analysis Group in 1984–1985 (16). The package provides the user with both macro-economic and energy sector models, with the former designed as a driver for the input side of the energy model. The two models can also been exercised independently. In fact, since the user is free to build whatever equation structure is deemed appropriate, ENERPLAN is again best described as a software package to build models, rather than a model per se.

The analytical approach for both maro-economic and energy sector modules can be summarized as follows:

1. Postulate an econometric model.
2. Enter the historical data for the variables in the program data base.
3. Conduct an econometric analysis to determine model coefficients.
4. Run the simulation model using the coefficients estimated in step 3.

Obviously, such an approach is extremely flexible, leaving the user to determine the level of detail at which he or she wishes to work. Indeed, as noted by the authors: ". . . the main strength of ENERPLAN is its programmability. Thus the program can handle models reflecting a whole range of schools of economic thought" (16, p. 155).

As part of the UNDP project, ENERPLAN was applied as a model testing exercise to two countries, Costa Rica and Thailand. The macro-economic model is compact, having just 14 equations. A key exogenous variable is the coffee price (included as a variable in the investment function), given the importance of coffee to foreign exchange earnings. Consumption, investment, imports, and exports are all based on simple econometrically estimated equations involving single period lags.

Obviously this is a very simple model, designed solely to serve as a demonstration. Indeed, at this level of simplicity, it is not clear that there is any advantage to having a model at all, since the only variable that is used for the subsequent energy analysis is GDP. This might just as well be defined in a set of completely exogenously specified scenarios.

This simplicity, however, should not obscure the potential value of the software package for use elsewhere. The package is extremely user-friendly, and the integration of the models with an econometric estimation package is excellent. There is, in fact, no reason why much more sophisticated models could not be developed with the software.

Typical output from the model can be transferred to a graphics module that generates the standard set of line, bar chart, and pie chart figures. The output side of the model appears to be limited and does not allow easy transfer to other standard packages for customized data and graphics manipulation. However, the necessary reprogramming effort to accomplish this appears small.

The ENERPLAN Energy Module

The energy model for Thailand contains some 60 equations, consisting of groups of equations for (1) final demand, (2) energy conversion, (3) primary energy supply, and (4) energy prices. Major exogenous variables (in fact determined by the macro-economic simulation) include real consumption (R.CP), the wholesale price index (WPI), and manufactured industry output (R.MANU). Energy demands use typical econometric equations of the type

$$Q(t) = a \cdot Q(t-1)^b \cdot Y^c \cdot P^d$$

where Y is income (or industrial activity)

P is price
Q is the demand at time t
a, b, c, d are estimated coefficients.

On the supply side, several variables are specified exogenously, such as the amount of hydropower and other domestic energy production levels. Note that all of the variables in the energy model are flows; there are no capacity equations. Imports are calculated as a residual. This is a fairly standard procedure for the construction of an energy model. A graphics option permits translation of tabular data to the usual graphical display formats.

Strengths and Limitations

Because the application reviewed here is in the nature of a field test for demonstration purposes rather than an actual model implementation by an energy planning agency, the potential strengths of the model for energy analysis may not be fully revealed. Nevertheless, a number of observations can be made.

First, the general programming environment is extremely user-friendly. The only major problem in this respect is the limitation in the number of characters allowable to identify each variable. The result is such variable names as ELES, ELEL, and OOLFN1, which makes interpretation somewhat tedious. New models could be implemented rather easily, which has advantages for assistance agencies that may need to implement many country models.

Second, the model is written around an existing econometric package. The user enters the data base, specifies the equations, and in a first step (choosing the menu entry "statistical analysis") can estimate the coefficients of the equations in question. Since this is perhaps a particularly suitable approach for estimating the coefficients of a macro-economic model, we see this as the main strength of ENERPLAN. As we note below, however, it is also one of the weaknesses with respect to the technical relationships in the energy sector.

From the standpoint of potential application to developing countries generally, the key shortcoming is the absence of a project level capability. Consequently there is also no relationship between energy project investment and output. For example, the output of coal + lignite = fired electric generation plants is given by the equation

$$HCEL = (.02203 - .00806*PCL/PCR$$
$$+ .97929*HCEL(-1)/(ELEL(-1)$$
$$- 0.3*HDEL(-1)$$
$$+ .14238*DUM79)*(ELEL - 3*HDEL)$$

where:

$$HCEL = \text{coal and lignite generation}$$
$$HDEL = \text{hydropower generation}$$
$$(-1) = \text{denotes time lag of one period}$$
$$ELEL = \text{total electricity}$$
$$DUM79 = \text{exogenously specified dummy variable}$$
$$PCL = \text{coal import price}$$
$$PCR = \text{crude import price.}$$

This is, of course, a typical equation in an econometrically specified model. It is unclear, however, how such an approach could be used to evaluate donor-financed projects (whether direct investments in generation capacity, or investments in transmission/distribution rehabilitation, or conservation projects, all of which would alter the dispatching of coal-fired power plants).

Again, given the very limited nature of the case study, it is perhaps unfair to criticize the model at such level of detail. However, the point is that any econometrically specified model of the electric sector will have great difficulty when evaluated at such a level of technical detail. The optimal dispatch process simply requires a completely different computational approach.

Clearly, since project-level detail is not included, it follows that energy sector investment and debt service requirements are also absent. Indeed, since the user manual makes no mention of how one might evaluate the macro-economic impacts of alternative projects or policies, it is fairly clear that ENERPLAN is not designed for such analysis.

THE TERI ENERGY PLANNERS SOFTWARE PACKAGE

The Tata Energy Research Institute (TERI) has developed a package of microcomputer models for use in energy planning (17). The package currently consists of five main modules:

Energy-economic simulation and evaluation model (TEESE).
Resource allocation model (an economic model of India)
Rural energy model
Energy conservation in buildings
Evaluation of renewable energy technologies.

THE TEESE MODEL

This energy-economy model embodies a number of recent hardware and software developments. It uses LOTUS-Symphony for data entry and matrix generation, and the commercially available SUNSET LP package, and it is the first BNL-type model in which the input-output is embedded in the LP as a constraint set. The LOTUS input format makes the model both flexible and user-friendly. The previous reliance on special matrix generation languages (such as the PDS/MAGEN code used in previous mainframe BNL models) is therefore eliminated.[4]

TEESE is suitable only for a rather broad strategic analysis. Lacking project-level

---

[4]The PDS/MAGEN software that was used in the mainframe era as a basis for many of the BNL and IEA models is a typical example of a very expensive mainframe software package that developing countries do well to avoid. Early versions in the late 1970s for CDC mainframes cost in excess of $40,000 to license.

detail, and with a very limited treatment of the electric sector, its strength is primarily in providing a macro-economically consistent view of energy demands and resources, and, through the shadow prices, some guidance on pricing policy. In fact, for large countries such as India, the lack of project-level detail is much less of an issue than for smaller countries, since any single project represents a much smaller perturbation to the overall system. Thus even in the electric sector, specification in terms of ranges of the capacity mix (in which the capacity variables in the LP model are simply subject to a bound set) is not that limiting for a national scale analysis for a country such as India.

Strengths and Limitations

Although the level of documentation is not as good as for some of the other models reviewed here, and some of the modules appear to be quite specific to India, the TERI models merit attention because they were developed not by consultants in a developed country, but by a leading research institution in a developing country. We have argued elsewhere that the programs and approaches advocated by "experts" from the developed countries have often proved unsatisfactory in the field. Here is a set of models that unquestionably reflect sound judgments about the suitability of particular approaches to the developing country context.

CONCLUSIONS AND RECOMMENDATIONS

This review of currently available software packages reveals significant differences in programming philosophy, assumptions about counterpart staff capabilities, and the contribution of models to the planning process. Moreover, it is clear that the path taken by each country is likely to be unique, given the large differences in institutional structure, size, technical capability, and planning philosophy.

Nevertheless, the following recommendations can be made, which should be used as a starting point for developing a new microcomputer-based energy analysis capability.

Commercially available packages like LOTUS 1-2-3 are the recommended programming environment for general analytical work. Existing RESGEN modules might be used as a starting point for staff training and the development of country-specific demand forecasting, and energy balance work.

The ENERPLAN package can be recommended for energy macro-economic modelling and econometric work. Its macro-economics module is well suited for estimating energy-economic relationships.

29

Software for Subsectoral Analysis

Although there are perhaps some tens of programs available for energy sectorwide planning, for subsectoral analysis the number of programs must number in the thousands. In part this is a reflection of the fact that at the subsectoral level—say for electricity utilities, refineries, and the like—many engineering, design, and financial calculations are routinely performed on microcomputers.

Moroever, with respect to potential application in developing countries, the issue of the suitability of software originally written for the developed countries is much less pressing than at the macro-economic or energy sectorwide level. This is because strictly technical analysis is largely independent of location: the laws of physics apply everywhere, and facilities such as refineries, central electric generating stations, and hydroelectric plants require much the same analysis in developing countries as in developed countries.

For example, what determines the equation structure of a linear programming model for a refinery is the engineering configuration rather than its location. Thus refinery modelling packages originally developed for the United States and Europe are much more readily applied in other parts of the world than are energy sectorwide models that must deal with the unique structure of developing country institutions and socioeconomic systems.

Because of the huge number of programs available, in this chapter we can provide only a representative sample. For each subsector we provide tabulations of packages known to the writers, and select one program in each group for detailed presentation. In the following sections, therefore, we present summaries for the subsectors typically important to developing countries: the fuelwood sector, new and renewable technologies, energy conservation, electricity, and the petroleum sector.

THE FUELWOOD SECTOR

The importance of the fuelwood sector to developing countries is well known and need not be reiterated here. Because of the widespread problems of deforestation

Table 29.1
Fuelwood Models

Model	Developed by	· Applied in
FRAP	ED/I, Washington	Many countries in Africa
BMASS	IDEA, Inc.	Sri Lanka
BIOCUT	Oak Ridge National Laboratory	Liberia
LEAP	ESRG/Beijer	Kenya

and the potential substitution between fuelwood and petroleum products (especially 1pg and kerosene), analysis of fuelwood problems has become a routine part of energy planning activity. Table 29.1 lists some of the fuelwood models currently available.

The Forest Resources Analysis and Planning Model (FRAP) is one of the better known fuelwood models, having been used in numerous African countries, including Morocco, Gambia, Mali, and the Sudan. FRAP is a detailed simulation model that projects forest resource suppy and demand over a 20-year-period for up to 20 separate regions in a country. Twenty-two input tables accept data by region on forest classifications, average mean annual increments, and stand volumes by forest zone, wood recovery, forest fires, herding, managed forests, and plantations. On the demand side, it accepts rural and urban populations and growth rates, fuel mixes, energy devices, and efficiencies, final energy demands by end use, industrial and commercial energy consumption, nonenergy wood demands and growth rates, and demand elasticities, among others (18).

FRAP is written in LOTUS 1-2-3, and is linked to a related model INVEST (not to be confused with ENVEST described in Chapter 28) that allows a user to create and evaluate potential investments, including fuel substitution projects, supply measures (plantations, agro-forestry, forest management) and device efficiency improvements (kilns, stoves, etc). Up to 180 different combinations of investment projects by region can be created on file and any combination assembled in portfolios and evaluated to assess their impact.

NEW AND RENEWABLE ENERGY

The diversity of microcomputer applications to the analysis of new and renewable energy technologies (see Table 29.2) is reflected by the titles of some of the papers presented at the Sri Lanka Symposium on Microcomputers in Developing Countries:

Microcomputer Simulation of Mass Transfer in an Alcohol Solar Still (Thailand)
Hydropower Surveys with Microcomputers
The Use of Computers in a Research Project on Solar Radiation Estimation and
    Measurement (Sri Lanka)

Indeed, microcomputers have become a routine part of research projects evaluating

Table 29.2
Models for the Analysis of New and Renewable Energy Systems

Model	Developed by	Technologies
TEESE (renewable technology evaluation)	TERI, India	biogas windmill PV solar
Prefeasibility evaluation	IDB	mini-hydro
Technical design + evaluation	University of Minnesota	mini-hydro

new and renewable technologies (both in the lab and in demonstration projects in the field where they are tied directly to analog data loggers).

The model for evaluating new and renewable energy technologies that is part of the TERI Energy Planners Software Package is typical of the sort of software now available for the evaluation of new and renewable technologies (19). The program is designed to evaluate the economic and financial feasibility of a variety of technologies to substitute for a variety of conventional technologies.

The model calculates investment costs, operating costs for both the new and the conventional system, and then evaluates payback period, net present value, and internal rate of return for the substitution in question.

ENERGY CONSERVATION PROJECTS

Much of the software suitable for the economic and financial analysis of new and renewable technologies is, of course, also directly applicable to the evaluation of energy conservation projects, since the criteria used are identical (payback period, net present value, internal rate of return). In addition, there is a great deal of software available focused on the evaluation of individual engineering problems. Indeed, probably the bulk of the microcomputer applications currently in use takes the form not of commercially marketed "packages," but of spreadsheets designed and used by plant managers to monitor and evaluate energy performances (see Table 29.3).

Table 29.3
Models for Evaluating Energy Conservation Projects

Model	Developed by	Applied in
ECPIE (economic/financial)	Hagler-Bailly	Sri Lanka
FCAM (economic/financial)	IDEA, Inc.	Dominican Republic
Steamprop, Heatless, Pipeloss	University of Texas, (Library of Engineering Programs) Center for Energy Studies	

THE ELECTRIC SECTOR

Electric utility planning and analysis software is widely used, but not widely available. Much of the software has been developed by large consulting firms in the developed countries, and by large utilities in developing countries (such as PLN in Indonesia, or the Central Electric Authority in India), entities that have the necessary experience and resources. Indeed, it is probably true that few energy planning bodies in developing countries have the necessary personnel to effectively utilitze detailed electric sector planning software.

The WASP model is one of the most widely used models for electric sector planning and has been used in many developing countries. It was developed originally as a mainframe model under the auspices of the International Atomic Energy Commission, written in FORTRAN, and as soon as microcomputers became more powerful there have been numerous efforts to download WASP. As can be seen from Table 29.4, there are at least four different microcomputer versions of WASP.

The micro versions differ not only in their user interfaces, but also in terms of the substantive calculations and the extent to which the original WASP algorithms have been improved. For example, the ENTEK version pays special attention to the interaction with the grid of nondispatchable technologies (micro-hydro, wind, solar, etc.), and to marginal cost pricing issues (20), whereas the ANL version has been placed into a shell that also includes spreadsheet modules for demand forecasting (in a package known as ENPEP).

PETROLEUM SECTOR

The points made above with respect to the electric sector apply equally to the petroleum sector. In the oil exploration and development area, for example, there are numerous microcomputer software packages in routine use throughout the world. However, this is of such a specialized nature, requiring extensive experience and training, that is has little relevance to energy planning bodies. Almost all of such software deals with evaluation of seismic, aeromagnetic survey and drilling information. Table 29.5 lists a sample of such software.

Table 29.4
Models for Electricity Planning

Model	Developed by	Applied in
WASP (capacity expansion	Asian Development Bank PLN	Pakistan Indonesia
ISPLAN (generation/ transmission/fuel supply)	IDEA, Inc.	India Thailand
Load flow, transient analysis	ACRES	Tanzania

Table 29.5

Software for Petroleum Exploration and Development

Vendor	
Molli Computer Systems (Colorado Springs, CO)	Back pressure curve analysis Historical decline curve analysis oil + gas well forecasting
Petrocomp Systems (Houston, TX)	Land management, reserve reporting Production history + forecasting
Resource Engineering (Fullerton, CA)	Reservoir simulation, waterflood performance evaluation
Pennwell (Tulsa, OK)	Petroleum formation evaluation

A number of microcomputer refinery models have been built for developing countries, usually of a scale suitable for general planning purposes rather than day-to-day operational management (i.e., involving a few hundred rows and columns, rather than the thousands of rows and columns characteristic of mainframe models used for operational optimization by world-scale refineries).

The World Bank has recently developed a refinery LP for general use called RLPM (refinery linear programming model), which is based on a general refinery configuration. The program is written entirely in FORTRAN, using a standard editor to update the data files. Although the model may not have sufficient flexibility to model precisely the configuration of a particular refinery, it is most suitable as a training tool.

30

Implementing Microcomputer-Based Energy Planning

THE HARDWARE DECISION

As noted in Chapter 26, one of the most fundamental impacts of microcomputers has been to reverse the traditional order of priorities. In an era when mainframe computers required the outlay of hundreds of thousands of dollars, it was natural that the hardware acquisition decision assumed great importance. Today the cost of microcomputers is very small, not just in absolute terms, but also when compared to the cost of expatriate experts that characterize the typical technical assistance project.

In fact, the cost of microcomputers is currently so low that it should be viewed not as a capital expense, but as an operating expense. Indeed, with the pace of technological changes as rapid as it is, one would expect to replace such machines every few years.

With the rapid advances in technology has also come a blurring of the distinction between microcomputers and mini-computers, with so-called work-stations (from vendors such as Apollo) positioned somewhere in-between. Certainly for most energy planning activities in developing countries, the requisite hardware requirements are at the "high" end of the microcomputer range.

Moreover, there is little question that the de facto operating system standard for energy, macro-economic, and technical modelling, throughout the world is MS-DOS. All of the models presented at the 1985 United Nations/DTCD Symposium on Microcomputer Modelling were based on MS-DOS, running under PC-DOS, or MS-DOS-based personal computers. The same is true of all of the models presented at the 1984 Symposium on Microcomputers in Development in Sri Lanka, and even in the larger countries such as India, where until recently the domestic microcomputer industry appeared to be concentrating on 8-bit, CPM-based machines, over the last year numerous MS-DOS-capable machines.

The main reason for MS-DOS-based machines is not necessarily the intrinsic superiority of this standard, but one of software availability. Not only can one benefit from the energy models developed for other countries, but also from the vast number of commercial software packages designed expressly for the sort of analytical activity that is characteristic of energy planning.

Nevertheless, even given the need for MS-DOS-based systems, there remain numerous issues to be faced by those contemplating microcomputer-based energy planning. Where should the hardware be bought? How can one be assured of adequate maintenance support?

LOCAL PURCHASE?

In most countries of the Eastern hemisphere, the power supply is 220 volt, 50 Hz, as opposed to the Western hemisphere standard of 110 volt, 60 Hz. In Africa and Asia, therefore, local rather than the U.S. purchase is often advocated on ground of power supply suitability. However, many machines now have built-in dual voltage capability. And in cases where a large number of machines are bought in the United States, the costs of a separate wiring at 110 volts with a single central power supply unit may be warranted. In any event, with most electro-mechanical devices such as disk drives operating on direct current, the only potential problem with simple (and cheap) step-down transformers is the degree to which 60 Hz computer power unit transformers themselves function reliably at 50 Hz (see the discussion on Morocco in Chapter 32).

Voltage stabilization equipment is recommended in almost all situations, irrespective of whether the machines are bought locally or in the United States. In some cases, uninterruptible power supply units may also be warranted where outages are frequent and prolonged. However, these are substantially more expensive than voltage stabilization equipment and may become less important as portable and lap-top microcomputers with a 4–5-hour battery life capture a larger share of the market.

A second argument for local purchase is maintenance. Machines bought locally can be serviced locally. In particular, it is sometime said that there is no incentive for a local company to provide maintenance on machines that were brought in by someone else. This may have been valid in 1983–1984, in the early phases of the rapid introduction of micros into developing countries. Today, however, even machines bought outside the country can be adequately serviced in most places. In any event, with the cost of hardware continuing to decline, maintenance becomes less and less of an issue. Broken parts are simply discarded and replaced by new units.

Such issues, however, are relatively minor compared to the actual hardware cost. Dealer markups are very high in most developing countries, with costs typically two to three times U.S. prices. Table 30.1 shows a price comparison of local dealer quotes in Rabat, Morocco, with U.S. prices (in early 1986). Indeed, in this particular case the price differentials were judged to be so great as to warrant purchase in the United States, even given the added costs of shipping and insurance. It should be noted that this price comparison is exclusive of import duties, which are not levied in the case of most technical assistance projects involving government agencies.

Table 30.1
Price Comparisons

	Rabat price, in $U.S.	ATT 6300 plus	IBM XT	PC compatible
IBM XT, 10Mb disk, 1 360K floppy	3915			
Memory extension to 512K	491 498	2595[†]	2395[‡]	1595[§]
DOS 2.1	69			
Keyboard AZERTY	268			
Monochrome screen	269			
256 to 640K memory extension	—	—	259.95[‖]	259.95
Graphics board	359	—	130.00[#]	130.00[#]
8087 Coprocesser	314	129	129	129
160 cps Epsom printer[‡]	753	369	369	369
Printer cable	59	30	30	30
	6995	3123	3312	2512

[*]As per exhibit.
[†]Comes with 640K installed.
[‡]Assumed to be the FX-85.
[§]Equivalent to IBM XT, but of non-IBM origin.
[‖]AST 6 pack plus 348K.
[#]Hercules compatible.

It is clear from recent experience that the hardware acquisition decision needs to be carefully examined on a country-by-country basis. Even if it is true that hardware costs are now quite low, an inappropriate decision at the outset of the process of implementing microcomputer-based planning may have serious consequences later on. Therefore an expert with wide experience should make, in the case of a technical assistance project perhaps as part of an inception mission, a careful assessment of the advantages and disadvantages of each option.

THE SOFTWARE ENVIRONMENT

With respect to programming software, similar de facto standards have emerged. For spreadsheet work, the leading software package is still LOTUS 1-2-3. To be sure, there are other spreadsheets (such as FRAMEWORK, Symphony, and Supercalc) that have seen application for some models (such as the FRAMEWORK-based model in Costa Rica). But as in the case of the desirability of MS-DOS compatibility, the issue is not one of intrinsic superiority but of cost-effectiveness.

Where compiled languages are used, for the moment FORTRAN, Pascal, and BASIC remain standard, particularly in the versions developed expressly for micro-computers, such as TURBO Pascal and BASIC (by Borland) and QUICK-BASIC (by Microsoft). More advanced languages such as APL and PL/I are not likely to be useful in developing country situations where energy planning agency staff are largely unfamiliar with these languages.

A SOFTWARE CLASSIFICATION

It becomes apparent that from a software standpoint the energy modelling frameworks fall into a number of quite distinct groups.

1. Compiled programming environments, using high level languages such as FORTRAN and Pascal, whose programming structure is unique to the country in question (such as the ENVEST model for Morocco or the ENERSTAT model developed for the General Petroleum Corporation of the Sudan).
2. Compiled programming environments, also using high level languages, but whose programming environment has been structured in such a way as to allow the user considerable flexibility in equation structure (such as the ENERPLAN model).
3. Pure spreadsheet environments, based exclusively in commercial spreadsheet packages (such as Morocco and Pakistan RESGEN).
4. Hybrid environments, consisting usually of combinations of LOTUS-based spreadsheets and compiled code for specific algorithms (such as electric sector optimal dispatch or refinery optimization) whose implementation in a compiled language is more efficient (such as the TERI TEESE model and the Morocco refinery/petroleum sector model, which use LOTUS as the data interface in conjunction with a commercially available LP optimizer).

From the standpoint of most national energy planning institutions and the donor community, where resources do not allow reprogramming of a complied code for every new country, the first group of models is clearly unsuitable, even though some include a number of otherwise useful attributes (such as the ENVEST project analysis). But even among groups 2–4 there exist a number of significant differences from the standpoint of suitability for general application.

The models of group 2 have a number of desirable features from a software perspective. As noted in Chapter 28, ENERPLAN, for example, is extremely flexible, allowing the user to specify almost any simultaneous equation structure for both the macro-economic simulation, the forward linkage, and the energy model. The model is user-friendly and very easy to operate.

However, many of the models in this group do not have the ability to incorporate project-level information, which we have argued is a critical requirement for energy investment planning. There is, of course, no particular reason why the programs involved might not be extended to incorporate this requirement, but it seems unlikely that the effort would be worthwhile given that the group 3 and 4 models already permit this.

The most serious deficiency of most compiled programs, however, is the lack of attention paid to interfacing with other programs, in particular with currently available spreadsheet programs.

The flexibility of spreadsheets, the ease of updating and modification argue strongly for LOTUS-based models. One important advantage is that many subsectoral models are now also being written in commercially available spreadsheets, which can easily be integrated. For example, the new Morocco energy model (based on LOTUS RES-GEN) has a direct link to the FRAP model, a LOTUS-based model for the fuelwood sector that has been widely used in Africa.

A further argument for the use of commercially available spreadsheets concerns presentational aspects. Presentation of results to decision makers mandates professional quality graphics for effective communication.

Despite all of these advantages, however, one must inject a certain note of caution. First there seems to be a natural tendency to try to do everything in spreadsheet packages, even though it is clear that in some cases this is quite inefficient. There is no doubt that an algorithm for the optimal dispatch of electric generation units into a load-duration curve is much more efficient when done in a FORTRAN or PASCAL environment. The same holds true for even small linear programs, for which there are now several excellent microcomputer optimizers.

CONCLUSIONS

In sum, the most useful programming environment for general application is likely to be based on spreadsheet packages, complemented by appropriate packages for statistical analysis, and linear programming. The programming environment should also make maximum use of the numerous enhancement products and utilities now available for spreadsheet documentation, graphics enhancements, window generators, and the like.

In terms of hardware, no model likely to be useful for energy planning is likely to face a hardware constraint, beyond the need for a hard disk, and the usual 640k of RAM. Very large spreadsheets, which with an add-on extended memory board can now be up to 2MB in size, should probably be avoided on efficiency grounds. Where larger LP models are being considered, an 8 Mhz MS-DOS type machine, accelerated with an 80287 math chip, can bring computation times for LPs even of several hundred rows and columns down to a few minutes.

31

Potential Constraints and Limitations

The decreasing cost of the microcomputer and the availability of a fast-growing body of software suitable for energy analysis have resulted in a very rapid dissemination of microcomputers in developing countries in general, and among energy planning entities in particular. As noted earlier, microcomputers inherently address many of the problems encountered in the application of mainframes and minis to energy planning problems and bring to energy planning agencies an extremely powerful capability.

Nevertheless, there are a number of cautions that must be noted. Gifford (21) identifies four main issues: the black box syndrome, infinite accuracy, data/software integrity, and instant expertise. Whereas these issues are to some extent characteristic of all computers, mainframes as well as micros, the ready availability of microcomputers in developing countries requires some special cautions.

THE BLACK BOX SYNDROME

As software packages become more complex, the black box problem becomes more acute. Indeed, with the rapid advances in microcomputer hardware, the potential for models becoming needlessly complex also increases. In the late 1970s large-scale energy system LPs became ever larger as hardware manufacturers developed super-efficient LP solvers that could solve models of thousands of constraints: the MAR-KAL model implementation for the IEA had several thousand variables and rows, yet could be solved on large mainframes in a few minutes.

One of the major advantages of the spreadsheet approach to energy modelling is its complete openness. All assumptions and formulae are readily available to the user and can be verified rather easily. Spreadsheet models can be described as "glass boxes" rather than "black boxes."

INFINITE ACCURACY

A recent discussion concerning the error introduced by a six-block linearization of a (continuous) annual load-duration curve for a developing country illustrates this problem. The engineers pointed to the dangers that the approximation errors would distort a subsequent fuel consumption calculation. Yet a proposer perspective indicates that any such approximation error is at least one order of magnitude smaller than the uncertainties that underlie the demand projection itself (which defines the total area under the curve).

DATA/SOFTWARE INTEGRITY

The "garbage in, garbage out" phenomenon that has long plagued computerization has perhaps become less of an issue with modern microcomputers given their user-friendliness and the ease with which data input screens with built-in error checking can be built. At the same time, however, new difficulties have arisen with respect to the verification of spreadsheets. There are many microcomputer experts who claim that any spreadsheet of any size is likely to contain at least one error. The need for adequate documentation and verification of spreadsheets cannot be overemphasized.

INSTANT EXPERTISE

Orenstein (22) puts the problem this way: "Anyone having the purchase price can now acquire advanced capabilities and instant expertise. The great danger is not that bad software will be acquired: the danger is that the engineer who purchases the software will not know the difference. The danger is that good and powerful software in the hands of an inexperienced engineer may become life-threatening, and the danger is also that a good and competent engineer will be too easily tempted to step into unfamiliar and dangerous waters."

Whereas this was written in the context of software suitable for the design of civil engineering structures (hence the reference to "life-threatening" consequences), the general admonition remains entirely valid to the use of microcomputers in developing countries for planning purposes. One of the biggest dangers concerns the ability to differentiate between model results that are counterintuitive but correct, and those that are counterintuitive because of mistakes in either the data input or in the specification of equations. Only an individual who is well trained and has significant technical experience is in a position to make such judgment calls, particularly when dealing with complex models, or those with substantial technical content. Yet the shortage of experienced, well-trained individuals in developing country energy planning entities is the binding constraint on energy planning activity in general.

32

Case Studies

In this chapter we present three very brief case studies designed to illustrate some of the points made in earlier chapters in the context of real energy planning projects.

SRI LANKA

The Sri Lanka microcomputer-based energy modelling system was developed in 1984–1985 for the Ministry of Power and Energy under the auspices of an ADB technical assistance project in support of the development of a national energy strategy. It was the first microcomputer modelling system to be designed as a hierarchial system closely aligned to the realities of planning in Sri lanka. Figure 32.1 illustrates the general structure.

At the first level of the hierarchy is the energy macro-economic accounting system. This reflects the input to the overall planning process of the Ministry of Finance and Planning. Written in LOTUS 1-2-3, it communicates with the next level of the hierarchy by a set of ASCII interface files: GDP, exchange rates, world oil price scenarios, and the like are passed down to the energy models (where it is used to generate macro-economically consistent demand projections), whereas energy sector investment, debt service, and energy imports and exports are passed up from the energy model.

The RESGEN-3 and EFAM-1 models are used at the second level of the hierarchy, which reflects the aggregate energy sector. RESGEN-3 is an RES-based model that was reviewed in Chapter 28. EFAM (for energy finance assessment model) provides the institution-specific financial flows corresponding to the RESGEN energy flow. The model multiplies the physical energy flows of RESGEN by the price structure to yield the corresponding financial transactions.

At the third level of the hierarchy is a series of subsectoral models. A linear programming model of the Colombo refinery was developed jointly by the energy planning group and the refinery, which provides RESGEN with details on refinery configuration. This model is somewhat more complex than the Systems Europe model

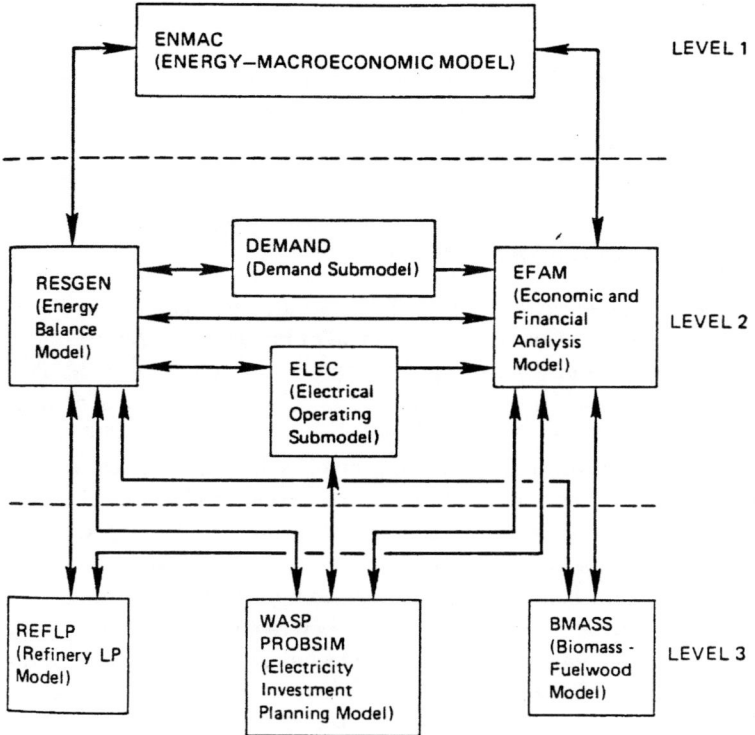

Figure 32.1 The structure of the Sri Lanka models.

for Thailand (see Chapter 26), simulating every major unit operation of the refinery, and its many blend streams.

RESGEN was designed to use directly the output of electric sector capacity expansion plans as they are typically prepared by electric utilities. In the Sri Lanka scheme, this transfer is accomplished by a member of the Ceylon Electric Board Planning unit staff who is involved in running both RESGEN and the detailed electricity planning models run by the CEB.

Noncommercial fuels, primarily fuelwood and agricultural wastes, are examined in a separate model, with its aggregate outputs fed manually to the energy sector models. Written originally in FORTRAN, it was designed specifically to examine the implications of alternative cookstove and fuelwood plantation programs.

Hardware Issues

Machines were acquired locally: a WANG PC with a 20Mb hard disk, and a Tandy 2000 with high-resolution color graphics. There were considerable doubts about the IBM-PC compatibility of the WANG PC, but these were resolved in early 1984 by the introduction of a PC emulation board.

The hardware solicitation process was difficult, with many claims made by local vendors about the availability of software that proved to be inaccurate. The planning project therefore started by implementing a simple version of RESGEN on an old Radio Shack Model II that was available in the ministry. Although the software later had to be transferred to the MS-DOS equipment, this strategy had the advantage that training and data collection could begin at the earliest phases of the project.

Lessons of the Sri Lanka Experience

The Sri Lanka experience was in general successful, particularly from the standpoint of providing useful input into the development of the national energy strategy. However, it is clear that the level of staff training was untypically good and that the institutional structure was untypically favorable. This permitted an approach that was fairly sophisticated. The modelling group included experienced individuals seconded from the subsectoral institutions who are involved in the technical details of subsectoral planning and who played key roles in developing the model structure.

For example, the refinery model equation structure was developed largely by the refinery engineer, with the consultant providing only the overall guidance. As a result, the model is now being run and maintained by the refinery planning group. The Sri Lanka experience thus illustrates another key requirement for a successful modelling effort: only individuals with adequate technical training are able to distinguish model results that are counterintuitive but correct, and results that are counterintuitive but incorrect due to data errors or equation specification problems.

Once the initial acquisition problems were overcome, the hardware arrangements have given few difficulties. Maintenance by local companies has been generally timely. The WANG equipment has been especially robust, surviving at least one major failure of the voltage stabilization equipment without ill effects to the motherboard. The higher first costs of buying all of the equipment locally has proven to be the correct decision in this case.

MOROCCO

Until 1985 the Service de Planification et Documentation of the Ministry of Energy and Mines used the ENVEST model, reviewed in Chapter 28. However, as part of a hardware modernization program that replaced the Apple-IIs with IBM and Compag ATs, ENVEST is now being replaced with a hierarchial RESGEN-type system in a LOTUS programming environment.

From the standpoint of both the hardware environment (8 Mhz ATs with Bernoulli box/hard disk storage), and the enhanced Lotus programming environment, the Morocco application is a good reflection of the state-of-the-art of microcomputer modelling. The list of hardware installed in the first three months of the project is detailed in Table 32.1.

The project also makes extensive use of commercial packages for graphics, spread-

Table 32.1
Hardware Inventory in Moroco

1	IBM AT, with 640k, 20Mb hard disk, Bernoulli box, 1 MB expanded memory, IBM proprinter (dot matrix) Used as the main machine for modelling and data base work.
1	COMPAQ-II (8Mhz, 80286 based portable, with two 360k floppy disk drives) Used in the field by visiting consultants in hotel rooms(!) and for out-of-office demos.
1	IBM XT, 640k, 20MB hard disk, IBM EGA enhanced color graphics adaptor, Hewlett Packard six-pen plotter Used for presentation quality graphics.
2	IBM PCs, 640k, twin floppy drives, one with IBM Wheelwriter letter quality printer, one with EPSON wide carriage dot matrix printer.

sheet verification and documentation, and data base manipulation. Table 32.2 lists the commercial software packages that were acquired in the first year of the proejct.

Lessons of the Morocco Experience

In light of the discussion in Chapter 30, a decision was made to purchase most of the hardware in the United States because of the substantial cost savings (see Table 30.1). The only exceptions were the French keyboards, which were purchased locally (and used for word processing).

The power supply gave a number of problems, with 50% of the supply transformers needing replacement in the first 12 months of the project. At first it was thought that these difficulties stemmed from 50 Hz rather than 60 Hz supply, but the manufacturers insisted that this would not be a source of trouble. The cause was finally traced to overvoltages resulting from incorrect transformer settings in the uninterruptible power supplies (UPS).

The modelling system is currently being used to evaluate projects submitted for the next five-year plan by the parastatals over which the ministry has jurisdiction, and to support a petroleum pricing study. The first tangible product of the new mi-

Table 32.2
Commercial Software Used in Morocco

Software	Function
LOTUS 1-2-3, version 2	Basic energy model
FREELANCE	presentation quality graphics
WORDSTAR (French version)	Word processing
ATLAS	mapping program to display regional data
LP83	linear programming package for refinery model
SQZ	utility to reduce storage requirements of LOTUS files
IBM hard disk organizer	Utility
STATPLAN	Statistical package (used, e.g., for estimating demand model elasticities)

crocomputer support system has been the production of the 1985 statistical report on the petroleum subsector.

THE SUDAN

Energy sector planning in the Sudan is the responsibility of the National Energy Administration (NEA), which acquired its first microcomputer in early 1982 (see Table 32.3). Unlike the Sri Lanka and Morocco examples, where microcomputers were introduced as part of technical assistance projects in support of very specific analysis and decision-making needs, in the Sudan the introduction of microcomputers evolved almost spontaneously in the appearance of a 64k Ohio Scientific machine in late 1982, left there by an unrelated project.

The Ohio Scientific machine quickly became inadequate as the analytical demands increased. NEA began to use Northstar microcomputers owned by the Ministry of Agriculture during evenings, and the University of Khartoum's mainframe, but the need for dedicated equipment at NEA became pressing. At this time the Georgia Institute of Technology was using an Osborne for evaluation of renewable energy projects (running Supercalc software), and thus NEA, too, acquired an Osborne portable in early 1983. This was followed in July by the first IBM PC.

Further IBMs have been acquired since then, and by 1985 the Ohio, Osborne, and Hyperion machines were withdrawn from service due to lack of spare parts and inability to provide maintenance. LOTUS 1-2-3 was introduced in mid-1984, and has become the main software used for analytical work. Table 32.4 shows the commercial software in use at NEA.

The NEA computer unit has collected some interesting statistics on how microcomputers are really used. In 1986 this sample data shows the following breakdown (as percentage of total hours of microcomputer use):

Quantitative analysis	25%
Data base management	27%
Word processing	19.6%
Training	26.1%
Preparation of graphics	0.8%
Programming	1.4%

Statistics show that only 1% of total possible machine hours were lost to breakdowns and power cuts. The data also show the following patterns of use by the different sections of NEA, a fairly good proxy for the type of technical work undertaken:

National energy plan	6.7%
Biomass	1.7%
Pricing	6.7%
Information	9.8%
Regional planning	0.9%

Table 32.3
Microcomputers in the NEA

Description	RAM	Storage	Introduced
Ohio Scientific	64k	2DD	early 1982
Osborn Portable	256k	2DD	late 1982
Hyperion Portable	256k	2DD	early 1983
Hyperion Portable	256k	2DD	early 1984
IBM PC Monochrome	512k	2DD	late 1984
IBM XT, color	640k	10MbHD	early 1985
IBM XT, color	640k	10MbHD	late 1985
IBM XT, color	640k	10MbHD	early 1986
COMPAQ portable	640k	10Mb	early 1986

Table 32.4
Commercial Software Used by NEA

Software	Acquired	Function
1-2-3 DBASEIII	1984	General programming
EASYWRITER	1983	
MULTIMATE	1986	Word processing
1-2-3 REPORTWRITER	1986	
VOLKSWRITER	1985	
LINDO	1985	Linear programming
NORTON UTILITIES	1985	File management
BASIC	1983	
QUICKBASIC	1986	General programming
STATPLAN	1983	Statistical analysis

Computer	42.0%
Energy planning project	3.9%
Projects	29.0%
Evening	1.8%

The high proportion of use by the computer section reflects its involvement in training NEA staff as a whole, some 85% of whom received instruction by the unit in 1986.

Lessons of the Sudan Experience

The Sudan experience shows that a formal energy planning package (of the type reviewed in Chapter 28) is not necessarily essential to a successful application of microcomputers to energy planning. Indeed, an analytical system developed by the local staff has a number of advantages, not the least of which is the high likelihood that it will in fact be used. To be sure, as staff capabilities improve, a more formal system might prove useful, but certainly the present system seems well matched to the capabilities of the NEA staff.

The climate and power supply environment in Khartoum must be viewed as extremely hostile, with extremely high temperatures (albeit under low humidity), strong ambient temperature variations, high dust levels, and frequent power outages and abnormal voltage and frequency transients being commonplace. Moreover, given the generally difficult economic conditions, maintenance problems might be expected to be severe. Yet experience has shown that proper attention to the physical operating environment, and standardization on IBM and name brand MS-DOS-based systems, can successfully overcome such potential difficulties.

33

General Conclusions

Microcomputers have the potential for making a significant contribution to energy planning in developing countries. They are ideally suited to the computational and data needs of energy analysis, and they ease the process of communication with the decision makers.

At the same time, it should be noted that microcomputers do little to change the fundamental issues that beset energy planning in developing countries: leadership and management of energy sector institutions, adequacy of resources, and staff capabilities and training. Indeed, microcomputers can make a direct contribution only to the last of these three areas, by virtue of their suitability as an adjunct to training programs. Without improved leadership and management of the agencies involved, however, microcomputers will become little more than toys, and contribute little to the actual decision-making process.

Given these realities, the technical assistance community has a certain responsibility to go beyond just the initial granting of funds to acquire equipment and software, but to provide some on-going institutional support. The best example of the sort of commitment that is necessary to support computer-based energy planning is the support given by the International Atomic Energy Commission (IAEA) to the WASP model and electric utility planning. Whereas the IAEA does not usually provide any hardware, training programs (at Argonne National Laboratory) have been offered for many years to the point at which WASP is perhaps the most widely used planning electric utility planning model in developing countries. Moreover, IAEA provides on-going software support, with improved versions issued from time to time. Indeed, whatever the shortcomings of WASP, the fact remains that IAEA and the use of WASP has made a substantial contribution to the improvement of electric utility planning procedures that has measureably improved the quality of investment decisions.

Similar action has been initiated by United Nations/DTCD with ENERPLAN, which has been distributed to about 30 government agencies and nonprofit organizations with interest in energy planning. The version of ENERPLAN has been upgraded and a new substantially modified version, ENERPLAN 2, is in the process of preparation.

There is also a need for better communication among developing countries. There is relatively little exchange of information and experience among those currently using microcomputers for energy planning. Most of the information comes to countries filtered through the perspective of expert consultants from the United States and Europe, rather than from independent sources in other Third World countries. UNDP might do more to provide some centralized service for the exchange of information, advice, and software.

Beyond the relatively straightforward cautions about the need for adequate power supply protection, microcomputer hardware issues themselves are relatively straightforward. However, the issues surrounding maintenance and optimal acquisition modalities do need adequate attention at the earliest point of the implementation process. This is best accomplished by an inception mission by an expert with broad experience in installing similar systems in other countries.

Given these many software packages for energy planning, what can be recommended for general use? Indeed, as the above review suggests, there is no single piece of software that can serve all energy analysis needs of a developing country's energy agency. As noted in the case studies, a mix of software has been introduced in all of the countries in which microcomputers have been installed. Even though there continue to be new energy planning software packages developed (largely by universities and national laboratories in Europe and the United States), in fact, that goal is likely to remain elusive.

References to Part IV

1. M. Munasinghe, M. Dow, and J. Fritz. 1985. "Microcomputers and Development," National Academy of Science-CINTEC.

2. Natural Resources Division, Department of Technical Co-operation for Development, United Nations. 1985. Microcomputer Applications for Energy Planning in Developing Countries.

3. World Bank. 1979. Revised Minimum Standards Model, Economic Analysis and Projections Department.

4. World Bank. 1985. "Assessment of Electric Power System Planning Models," Energy Department Paper no. 20, Washington, DC.

5. R. Hutchinson, S. McBrien, and M. Waller. 1981. "Preliminary Survey and Selection of Computer Models Suitable for Energy Planning Applications in Developing Countries," MITRE report MTR 81 W205 to the Commission of European Communities, ISPRA.

6. Kydes, A. 1978. "The Brookhaven Energy System Optimization Model," Brookhaven National Laboratory, BNL 50873.

7. Systems Europe S.A. 1986. "Energy Systems Analysis and Planning," Doc 851.12-B, May, Bruxelles.

8. Garza, F. and A. Manne. 1973. "ENERGETICOS: A Process Model of the Energy sectors," in Goreaux and Manne, eds.: *Multi-Level Planning: Case Studies in Mexico,* North-Holland. Amsterdam.

9. Fishbone Y., and H. Abilock. "MARKAL: A Linear Programming Model for Energy Systems Analysis: Technical description of the BNL version," *Energy research,* 5: 353–357.

10. Nathan Robert R., et al., 1982. "Thailand Energy Master Plan Project," report to NEA/ADB/UNDP.

11. Gordian Associates. 1981. "A Linear Programming Model of the Tunisian Energy System," report to l'Enterprise Tunisienne d'Activites Petrolieres (ETAP).

12. Malone R., and A. Reisman. 1978. "Less-Developed Countries Energy System Network Simulator LDC-ESNS," BNL 50836 Brookhaven National Laboratory, BNL 50836.

13. Derrico T., et al. 1982. "EDIS User Manual," Institute for Energy Research, State University of New York at Stony Brook.

14. Thailand Energy Master Plan Project, Report by Rober R. Nathan, Systems Europe et al to ADB/UNDP and the National Energy Administration of Thailand, 1982.

15. Gorden, M. 1985. "ENVEST: A Planning System for Energy Sector Investments," presented at the United Nations/DTCD Workshop on Microcomputer Software Applications for Energy Planning in Developing Countries, New York.

16. The Tokyo Energy Analysis Group. 1985. "ENERPLAN User Manual." United Nations Department of Technical Cooperation for Development.

17. TERI. 1987. "The Energy Planners Software Package," Delhi.

18. Forest Resource Analysis and Planning Model (FRAP), User Manual. 1986. report by EDI, Washington to Africa Bureau, USAID.

19. The TERI Energy Planner's Software Package. 1986. "Evaluation of Renewable Energy Technologies for End Uses." New Delhi: Tata Energy Research Institute.

20. Entek Research, Inc. 1986. "p/s Plan: A Software Package for Electric Utility Capacity Planning, Investment Evaluation and Rate Analysis," Setauket, NY.

21. Gifford J. 1987. "Microcomputers in Civil Engineering: Use and Misuse," *Journal of Computing in Civil Engineering,* Vol. 1, No. 1.

22. Orenstein, J. 1984. "Instant Expertise: A Danger of Small Computers," proceedings of the Third Conference on Computing in Civil Engineering, ASCE, pp. 578–582.